Cellular Interactions
and Immunobiology

BOOKS IN THE BIOTOL SERIES

The Molecular Fabric of Cells
Infrastructure and Activities of Cells

Techniques used in Bioproduct Analysis
Analysis of Amino Acids, Proteins and Nucleic Acids
Analysis of Carbohydrates and Lipids

Principles of Cell Energetics
Energy Sources for Cells
Biosynthesis and the Integration of Cell Metabolism

Genome Management in Prokaryotes
Genome Management in Eukaryotes

Crop Physiology
Crop Productivity

Functional Physiology
Cellular Interactions and Immunobiology
Defence Mechanisms

Bioprocess Technology: Modelling and Transport Phenomena
Operational Modes of Bioreactors

In vitro Cultivation of Micro-organisms
In vitro Cultivation of Plant Cells
In vitro Cultivation of Animal Cells

Bioreactor Design and Product Yield
Product Recovery in Bioprocess Technology

Techniques for Engineering Genes
Strategies for Engineering Organisms

Principles of Enzymology for Technological Applications
Technological Applications of Biocatalysts
Technological Applications of Immunochemicals

Biotechnological Innovations in Health Care

Biotechnological Innovations in Crop Improvement
Biotechnological Innovations in Animal Productivity

Biotechnological Innovations in Energy and Environmental Management

Biotechnological Innovations in Chemical Synthesis

Biotechnological Innovations in Food Processing

Biotechnology Source Book: Safety, Good Practice and Regulatory Affairs

BIOTOL

BIOTECHNOLOGY BY OPEN LEARNING

Cellular Interactions and Immunobiology

PUBLISHED ON BEHALF OF :

Open universiteit   and   **University of Greenwich (formerly Thames Polytechnic)**

Valkenburgerweg 167
6401 DL Heerlen
Nederland

Avery Hill Road
Eltham, London SE9 2HB
United Kingdom

BUTTERWORTH
HEINEMANN

Butterworth-Heinemann Ltd
Linacre House, Jordan Hill, Oxford OX2 8DP

ℛ A member of the Reed Elsevier plc group

OXFORD LONDON BOSTON
MUNICH NEW DELHI SINGAPORE SYDNEY
TOKYO TORONTO WELLINGTON

First published 1993
Reprinted 1994

© Butterworth-Heinemann Ltd 1993

British Library Cataloguing in Publication Data
A catalogue record for this book is
available from the British Library

Library of Congress Cataloguing in Publication Data
A catalogue record for this book is
available from the Library of Congress

ISBN 0 7506 0564 2

Composition by University of Greenwich
(formerly Thames Polytechnic)
Printed and Bound in Great Britain by
Martins the Printers Ltd.,
Berwick upon Tweed

The Biotol Project

The BIOTOL team

OPEN UNIVERSITEIT, THE NETHERLANDS
Prof M. C. E. van Dam-Mieras
Prof W. H. de Jeu
Prof J. de Vries

UNIVERSITY OF GREENWICH (FORMERLY THAMES POLYTECHNIC), UK
Prof B. R. Currell
Dr J. W. James
Dr C. K. Leach
Mr R. A. Patmore

This series of books has been developed through a collaboration between the Open universiteit of the Netherlands and University of Greenwich (formerly Thames Polytechnic) to provide a whole library of advanced level flexible learning materials including books, computer and video programmes. The series will be of particular value to those working in the chemical, pharmaceutical, health care, food and drinks, agriculture, and environmental, manufacturing and service industries. These industries will be increasingly faced with training problems as the use of biologically based techniques replaces or enhances chemical ones or indeed allows the development of products previously impossible.

The BIOTOL books may be studied privately, but specifically they provide a cost-effective major resource for in-house company training and are the basis for a wider range of courses (open, distance or traditional) from universities which, with practical and tutorial support, lead to recognised qualifications. There is a developing network of institutions throughout Europe to offer tutorial and practical support and courses based on BIOTOL both for those newly entering the field of biotechnology and for graduates looking for more advanced training. BIOTOL is for any one wishing to know about and use the principles and techniques of modern biotechnology whether they are technicians needing further education, new graduates wishing to extend their knowledge, mature staff faced with changing work or a new career, managers unfamiliar with the new technology or those returning to work after a career break.

Our learning texts, written in an informal and friendly style, embody the best characteristics of both open and distance learning to provide a flexible resource for individuals, training organisations, polytechnics and universities, and professional bodies. The content of each book has been carefully worked out between teachers and industry to lead students through a programme of work so that they may achieve clearly stated learning objectives. There are activities and exercises throughout the books, and self assessment questions that allow students to check their own progress and receive any necessary remedial help.

The books, within the series, are modular allowing students to select their own entry point depending on their knowledge and previous experience. These texts therefore remove the necessity for students to attend institution based lectures at specific times and places, bringing a new freedom to study their chosen subject at the time they need and a pace and place to suit them. This same freedom is highly beneficial to industry since staff can receive training without spending significant periods away from the workplace attending lectures and courses, and without altering work patterns.

Contributors

AUTHORS

Dr L. S. English, University of the West of England, Bristol, UK

EDITOR

Dr A. R. Waller, De Monfort University, Leicester, UK

SCIENTIFIC AND COURSE ADVISORS

Professor M. C. E. van Dam-Mieras, Open universiteit, Heerlen, The Netherlands

Dr C. K. Leach, De Monfort University, Leicester, UK

ACKNOWLEDGEMENTS

Grateful thanks are extended, not only to the authors, editors and course advisors, but to all those who have contributed to the development and production of this book. They include Mrs A. Allwright, Miss K. Brown, Miss J. Skelton and Professor R. Spier.

The development of this BIOTOL text has been funded by **COMETT, The European Community Action Programme for Education and Training for Technology**. Additional support was received from the Open universiteit of The Netherlands and by University of Greenwich (formerly Thames Polytechnic).

Contents

How to use an open learning text

An open learning text presents to you a very carefully thought out programme of study to achieve stated learning objectives, just as a lecturer does. Rather than just listening to a lecture once, and trying to make notes at the same time, you can with a BIOTOL text study it at your own pace, go back over bits you are unsure about and study wherever you choose. Of great importance are the self assessment questions (SAQs) which challenge your understanding and progress and the responses which provide some help if you have had difficulty. These SAQs are carefully thought out to check that you are indeed achieving the set objectives and therefore are a very important part of your study. Every so often in the text you will find the symbol Π, our open door to learning, which indicates an activity for you to do. You will probably find that this participation is a great help to learning so it is important not to skip it.

Whilst you can, as an open learner, study where and when you want, do try to find a place where you can work without disturbance. Most students aim to study a certain number of hours each day or each weekend. If you decide to study for several hours at once, take short breaks of five to ten minutes regularly as it helps to maintain a higher level of overall concentration.

Before you begin a detailed reading of the text, familiarise yourself with the general layout of the material. Have a look at the contents of the various chapters and flip through the pages to get a general impression of the way the subject is dealt with. Forget the old taboo of not writing in books. There is room for your comments, notes and answers; use it and make the book your own personal study record for future revision and reference.

At intervals you will find a summary and list of objectives. The summary will emphasise the important points covered by the material that you have read and the objectives will give you a check list of the things you should then be able to achieve. There are notes in the left hand margin, to help orientate you and emphasise new and important messages.

BIOTOL will be used by universities, polytechnics and colleges as well as industrial training organisations and professional bodies. The texts will form a basis for flexible courses of all types leading to certificates, diplomas and degrees often through credit accumulation and transfer arrangements. In future there will be additional resources available including videos and computer based training programmes.

Preface

Mammals have highly organised structures characterised by a large degree of cellular differentiation. To obtain food and oxygen and remove cellular waste products, the individual cells of the body depend on their interaction with the outside world via the circulatory system. Although of vital importance, the relation with the outside world also forms a potential threat to the organism as the external environment is also a source of harmful abiotic and biotic factors. Obviously the organism needs mechanisms of defence. The first lines of defence are the mechanical barriers between the individual and the environment; skin, gastrointestinal tract, urinary tract and lungs are part of this barrier. Defence against small chemicals resides in their detoxification in the liver followed by excretion in a faeces and urine, and defence against larger molecules and pathogens is in the hands of the immune system. The immune system is highly sophisticated and able to discriminate self from nonself. It also has a 'memory' of former encounters, making the response on a second encounter with the same enemy more efficient. The immune system is the subject of this BIOTOL book.

This book has been written with a minimal assumptions about the readers' previous knowledge of physiology, immunobiology and pathology. It is also equipped with a number of revisionary summaries and a glossary and the text incorporates many helpful reminders. These have enabled the author to produce a rigorous up-to-date treatment of this complex topic without losing the user-friendly house style of the BIOTOL series.

The first chapter gives a general introduction to immunology and introduces natural and innate immunity, adaptive immunity, the cell types and humoral factors involved in these and the importance of discriminating between self and nonself. It also provides an overview of the anatomy and function of the major lymph organs and of the lymph system.

Antibodies are the topic of the second chapter. Their general structural and functional characteristics are described first and subsequently the differences between antibody classes, antigen-antibody complex formation, opsonisation and further processing are described. Chapter 3 also deals with antibodies, this time focusing on the molecular genetics of antibody diversity. Chapter 4 places antibodies within the context of the defence reactions and describes the role of interacting T and B cells in their production. Discussion of cellular interactions automatically necessitates highlighting the importance of the Major Histocompatibility Complex, and description of antigen processing and presentation.

T and B cells are central in Chapter 5. This chapter starts with a description of T cell receptors and subsequently describes other cell surface molecules found on lymphoid cells. It then goes on to focus on the roles of T and B cells during development of the immune response. Chapter 6 continues the description of cellular interactions during the immune response. This time highlighting the roles of a variety of cytokines in this process. Chapter 7 describes the lymphatic system, the time course of an immune response, the different cell types participating and the importance of cellular adhesion molecules. The final part of Chapter 7 examines some of the possible negative consequences the immune system may have for the host. Cell mediated immunity is described in topic of the final chapter and here also possible negative consequences of the immune system for the host are dealt with.

This book on immunology has a functional orientation and clearly shows how amazingly sophisticated defence strategies are, both at the molecular level and at the level of co-operation within the system.

The author has combined an authoritative, up to date account of this rapidly advancing field of study, with a thought provoking interactive style. In combination these two attributes provide readers with an opportunity to grasp the key issues of immunobiology.

Understanding these issues opens the way to the use of this knowledge in many important and developing areas of application. These latter issues are dealt with in the BIOTOL text 'Technological Applications of Immunochemicals'.

Scientific and Course Advisors: Professor M. C. E. van Dam-Mieras
Dr C.K. Leach

Innate and acquired immunity

Innate and acquired immunity

In this introductory chapter, we will give you an overview of the immune system. We will begin by briefly looking at the historic development of the concept of immunity and the recognition of self and non-self. We will then examine two important aspects of immunity, the innate and acquired immunity and will explain in outline the mechanisms and limitations of each. We also provide an overview of the anatomy and function of lymphoid tissues and the important cells involved in the immune response. This will provide you with the framework in which to follow the discussions in subsequent chapters.

During this discussion we will introduce many of the terms used in immunology and you will learn of many of the important cell types involved in immunity. This overview will provide you with a framework in which to study the remaining chapters.

1.1 The historic concept of immunity

immunity

variolation

cowpox

We have to go back a long time in history to discover the probable beginnings of the ideas of immunity to infection. It is thought that as long ago as 500BC, Greek writers referred to immunity to a plague. The Chinese probably have the right to claim to be the first experimental immunologists in about 1000AD when they discovered that if they made children inhale powered scabs from patients recovering from smallpox it protected them from the disease ie the children became immune to smallpox. This led to the procedure known as variolation which was the introduction of smallpox crusts into the dermis of the skin which sometimes led to the death of the patient! Jenner is considered to be the father of modern immunobiology as he demonstrated that inoculation of cowpox crusts afforded protection to humans against smallpox. He had made the important observation that milkmaids who recovered from cowpox never contracted the more serious disease, smallpox; hence the name vaccination from the Latin name for a cow (Vacca). The milkmaids and the vaccinated individuals were protected from, or immune, to the virus.

1.2 Self and nonself

host

self
nonself

The basis of this protection was the ability of the immune systems of the milkmaids and vaccinated individuals (we will call the infected individual the host) to distinguish between self (their own tissues) and nonself (components of the smallpox virus). The host will raise a response to substances that are different to self components such as those found in bacteria, viruses, fungi and parasites which we collectively call pathogens and which cause infection in the host.

1.2.1 Plant pollens, venoms, drugs, red cells and transplants

immune
responses

antigenic

self

altered self

B lymphocytes
antibodies

There are many other nonself substances in the environment that also stimulate immune responses. These include plant pollens, various pollutants, toxins in our food, and venoms from biting insects and snakes. Many other substances that we use in our everyday lives may be recognised as nonself (antigenic) by our immune system. These include hair dyes and various cosmetics, drugs such as aspirin and many antibiotics. Some of these drugs bind to 'self' cellular proteins, such as those on red blood cells. The immune system sees these so called conjugated red cells as altered self and induces cells called B lymphocytes to produce antibodies which then bind to these altered cells and promote their destruction.

Π   Blood transfusions are recognised as nonself in the recipient. Can you explain why?

foreign

Transplanted organs are seen as foreign (nonself) by the recipient. As you may know, transplant patients are given immunosuppressive drugs throughout the lifetime of the transplant to prevent any destructive response by the recipient. This is because any one individual expresses a unique set of molecules on the surfaces of his/her cells which is slightly different to those on the cells of other individuals. The transfer of cells from one individual to another, as occurs in a transplant operation, results in recognition of these slight differences as nonself or foreign and this, in the absence of immunosuppressive drugs, would result in production of antibodies and activation of other immune mechanisms that would destroy the graft.

A similar situation holds for blood transfusions. Even though the blood has been matched for the main blood group antigens, the ABO system, the blood contains many other cells which carry the set of molecules unique to the blood donor's non-red cells. These would, therefore, be recognised as foreign by the recipient and these cells would be destroyed. The matched red cells would not be destroyed because these cells are identical to both donor and recipient and are considered to be self.

These nonself or foreign substances are called antigens. A bacterium is obviously a very complex antigen because it possesses many different proteins, carbohydrates and even lipids which are different to those of the host. We usually refer to each separate molecule, such as a protein, as an antigen, so a bacterium consists of many different antigens each of which can induce an immune response in the host.

acquired
immunity
adaptive
immunity

innate immunity

This ability to determine what is foreign to the host is the basis of what we call acquired or adaptive immunity. However, there are a number of other components or mechanisms which collectively provide protection to the host against pathogens without the need to recognise self or nonself. These are called innate immune mechanisms. Let us examine the basic features of these two types of immunity.

SAQ 1.1

Four of the following have a common feature; choose the odd one out.

1) Donor cells.

2) Sperm.

3) Horse serum.

4) Hen egg white lysozyme.

5) Host lymphocytes.

1.3 Natural and innate immunity

1.3.1 Physical barriers

exterior
defences
physical
barriers
skin
Sebaceous
glands

The healthy individual has a variety of ways to prevent invasion by a pathogen such as smallpox virus. We can divide innate defences into three main components; physical barriers, phagocytotic cells and soluble components (Figure 1.1). The first line of defence consists of exterior defences or physical barriers that prevent entry to the pathogen and make it very uncomfortable for the pathogen to even remain on the host. The most obvious is the skin which is a very effective barrier to most pathogens. Sebaceous glands in the skin also generate an acid environment by producing lactic acid which is antagonistic to many pathogens.

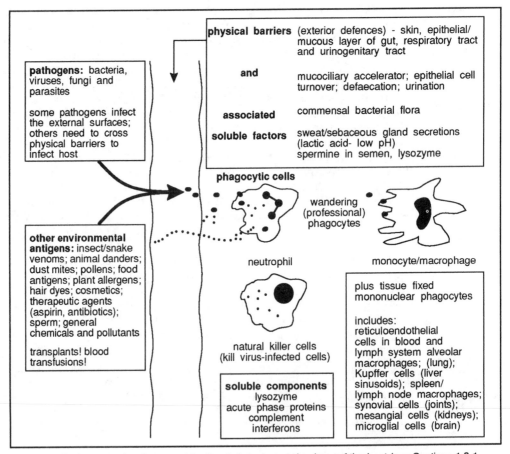

Figure 1.1 Environmental antigens and the innate immune mechanisms of the host (see Sections 1.3.1, 1.3.2 and 1.3.3 for details).

protective
mucous lining

The respiratory tract, the alimentary tract (the gut) and the urinogenitary tract have an exterior epithelial cell layer covered by a protective mucous lining. In the respiratory tract, cilia attached to the external surfaces of the epithelial cells continually beat upwards (mucociliary accelerator) towards the nasopharynx and this helps to expel particles and pathogens to the exterior. Epithelial cells are being constantly renewed and this sloughing off of old cells also expels pathogens lodged on the surface. The passing of faeces from the gut and urination are functions which also result in the clearing of unwanted pathogens from the body. Also, within the gut and vagina are commensal micro-organisms which prevent the colonisation of new invaders. Other body fluids contain molecules which are bactericidal (eg spermine in seminal fluid).

In spite of these defences, the majority of pathogens gain entry through the epithelial barrier although some pathogens (eg strains of *Streptococcus*) only infect the surface and do not attempt entry.

1.3.2 Phagocytic cells

If an organism manages to penetrate the physical barriers, it will encounter wandering cells (phagocytes) which can be summoned to the site of infection by a process known as chemotaxis. These cells engulf the pathogens or other particles and destroy them. This process is known as phagocytosis. The principal phagocytes are the neutrophils and macrophages. There are also fixed cells closely related to the mobile macrophages that are strategically placed in various sites throughout the body to intercept pathogens and other antigens; these cells are also strongly phagocytic. For instance, Kupffer cells are found lining the sinusoids of the liver through which the blood flows, reticuloendothelial cells line the insides of the blood vessels and lymphatics and alveolar macrophages are found in similar strategic positions in the lungs (Figure 1.1). A third type of cell, the Natural Killer cell, appears to be able to recognise cells infected with certain viruses such as Herpes and Influenza, and kill these cells releasing the viruses. The virus can then be destroyed by other means (see Chapter 8).

phagocytes
chemotaxis
phagocytosis

Kupffer cells

reticulo-endothelial cells

Natural Killer cell

1.3.3 Complement and other soluble components

In addition to these cellular components, there are various proteins in the body fluids which help to combat infection. Many secretions contain lysozyme which attacks the murein cell walls of bacteria. Some bacteria, viruses, and other antigens may activate a series of proteins collectively called complement which can not only kill the pathogens but also summon phagocytes to the inflammatory site to destroy them.

lysozyme

complement

In this process, complement proteins molecules called C3b are generated. These coat the nonself particles (viruses, bacteria). Some of the phagocytic cells have surface receptors for C3b and so the coated particles become attached to the phagocytotic cells. This process is called immunoadherence. The coated particles are then phagocytosed and destroyed. Particles which are coated by C3b proteins are said to be opsonised. The process of opsonisation increases the rate at which pathogens are phagocytosed. We will examine this process in greater detail in later chapters.

immuno-adherence

opsonisation

There are also a number of proteins called acute phase proteins whose concentration in serum can increase a 100-fold during infection. These proteins can promote the killing of various pathogens during infection. For instance, C-reactive protein binds to phosphoryl choline moieties on the surfaces of some bacteria such as *Pneumococci* promoting the activation of complement and subsequent killing of the bacteria.

acute phase proteins

C-reactive protein

Finally, a group of proteins called interferons, produced by virally infected cells, promote the synthesis of antiviral proteins by neighbouring cells, thus preventing the spread of the virus.

interferons

1.3.4 Important features of innate immunity

All of these physical barriers and associated soluble and cellular components make up the innate or natural immune system (Figure 1.2). It is so called because you are born with these mechanisms in place. Examine Figure 1.2 carefully as it provides a summary of the information given in Sections 1.3.2 and 1.3.3.

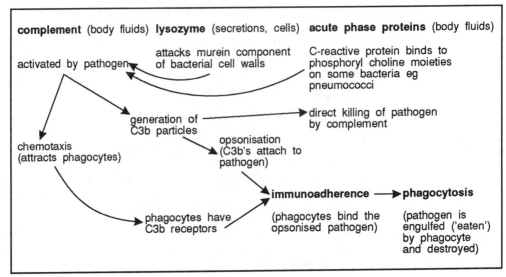

Figure 1.2 Some innate immune mechanisms.

There are two important features you need to know about the innate immune system.

Firstly, it is not antigen specific. This means that the cellular and soluble components cannot distinguish between self and nonself nor discriminate between one antigen and another.

Secondly, the resistance afforded by this system is not enhanced by repeated infections by the same pathogen. We say the system has no memory.

SAQ 1.2

Which of the following involve or result in the recognition of nonself?

1) Treatment of a diphtheria patient with horse antibodies to diphtheria toxin.

2) Killing of bacteria with spermidine in an individual.

3) C-reactive protein-mediated activation of complement.

4) A viral infection.

5) Plasma transfusion.

| **SAQ 1.3** | Which of the following are correct?

1) Interferons, lysozyme and C-reactive protein are antigen specific because they each bind to certain antigenic structures and not to any others.

2) Macrophages, natural killer cells, and neutrophils are effector cells in innate immunity.

3) Self proteins may be recognised as nonself when injected to another individual of the same species.

4) Serum proteins are not part of the innate immune system.

5) Bacteria must enter the body to be pathogenic.

1.3.5 Innate immunity is not enough

∏ Before you read on, see if you can think of any situation where innate immunity would not be effective in dealing with a pathogen.

The first problem with innate immunity is in its inability to recognise some antigens as being nonself. Although some soluble factors that have been mentioned kill some pathogens there may be difficulties encountered by the wandering phagocytes and fixed macrophages to decide what to phagocytose and what to leave alone when they cannot distinguish between self and nonself.

opsonisation

As was shown in Figure 1.2, there are mechanisms that promote engulfment and killing. As we have seen some bacterial species activate the complement system generating particles called C3b proteins which attach to and coat the bacteria, a process known as opsonisation (roughly interpreted it means 'make ready to eat'). Neutrophils and macrophages carry receptors on their surfaces for C3b's and these enable these cells to recognise and phagocytose opsonised bacteria.

Notice that the phagocytes do not need to recognise the bacteria as foreign. This is not, however, a universally effective mechanism as many bacteria, viruses, fungi, parasites and other antigens do not activate complement in this way.

bacterial adhesins

Recent evidence suggests that some bacteria express bacterial adhesins on their surfaces that can be bound by proteins on the surfaces of cells such as neutrophils. This may be another method by which phagocytes can attach to and subsequently destroy some species of bacteria. As with complement activation, however, there are many pathogens and antigens that do not possess such adhesins so this mechanism is only partially successful. An additional point to note here is that such adhesins are not inherently chemotactic ie they do not summon phagocytes, so these cells may never make contact with the pathogens anyway.

The result is that many extracellular pathogens and antigens successfully evade capture by phagocytes making innate immunity ineffective in many instances.

In addition to the inability to recognise self, innate immune mechanisms are totally ineffective when pathogens find their way into cells where they are not exposed to phagocytes or most of the soluble components of innate immunity.

humoral
immunity
cell mediated
immunity

These deficiencies can be corrected by the involvement of acquired or adaptive immune mechanisms resulting in the production of antigen specific antibodies (humoral immunity) or the activation of antigen specific cells (cell mediated immunity).

1.4 Acquired (adaptive) immunity - humoral immunity

Structural characteristics of antigens and immunogens

immunogens

epitopes

antigenic
determinants

Let us define antigens more clearly. We have already said that a complex pathogen such as a bacterium or virus consists of many individual molecules each of which can be called an antigen. If these individual antigens induce a response in the host then they are called immunogens. This response may result in the production of many different antibodies each of which will bind only to one area of some particular antigen and rarely to any other. We say that each antibody is specific. These distinct areas of antigen are called epitopes or antigenic determinants.

The major features of antigen structure are summarised for you in Figure 1.3 and indicate the different types of epitopes present in an antigen molecule.

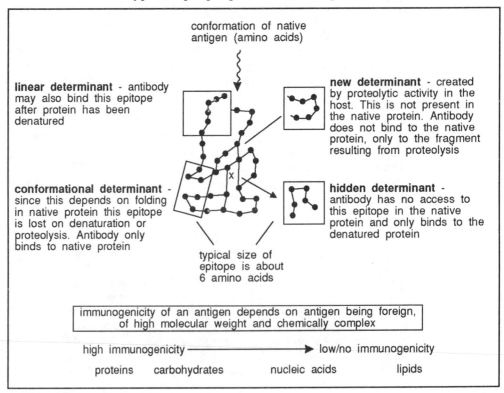

Figure 1.3 Major structural features of antigens (stylised).

∏ List the four types of epitopes we have displayed.

You should have identified linear determinants, conformational determinants, hidden determinants and new determinants. Note the features of each of these.

∏ What is the typical size of an epitope?

From Figure 1.3 you should have concluded that an epitope consists of about 6 amino acids.

∏ Which of the following is most effective at stimulating the production of antibodies? (carbohydrates, nucleic acids, proteins and lipids).

From the information given in Figure 1.3, the most immunogenic antigens are proteins and this is due to their foreignness, their high molecular weight and considerable chemical complexity. All foreign proteins invariably induce responses in normal healthy individuals.

Many protein types are common to a wide variety of species but they are generally immunogenic if transferred from one species to another. Consider serum albumin, a protein present in mammalian blood. Humans have human serum albumin (HSA) and mice have murine (mouse) serum albumin (MSA). It should be obvious to you that since both are called albumin they have basically the same structure and perform the same functions in Man and mouse. However, if you inject HSA into mice, the murine immune system will produce anti-HSA antibodies; the mouse has detected HSA to be nonself. Although HSA and MSA are largely homologous, the albumin of the two species does differ in amino sequence and some of these differences comprise epitopes to which antibodies will bind.

Carbohydrates are often immunogenic especially when attached to proteins as glycoproteins. Native DNA is rarely immunogenic although antibodies can be generated against polynucleotides especially when attached to proteins. Only a few anti-lipid antibodies have been reported.

SAQ 1.4

Relate items in left column to those in right column; use each item once only.

1) Conformational determinant   A) Cell walls

2) Phagocytosis   B) Amino acid sequence

3) Epitope   C) Folding

4) Linear determinant   D) Complement

5) Lysozyme   E) Antibody

1.4.2 The basic structure of antibodies

We shall examine the structure of antibodies in great detail in Chapter 2 but we will just give you some idea of their basic structure (Figure 1.4). There are many different forms of antibodies but the basic structure common to them all is that they are composed of two heavy chains (50 000 Daltons) and two light chains (25 000 Daltons). There are two antigen binding sites per molecule, each site is formed by the interaction of the amino acid sequence at the N-terminal end of one heavy chain and one light chain. The binding site can accommodate a hexasaccharide or hexapeptide. The average size of an epitope, then, is about six amino acids. Antibodies are produced by so called B cells. The body produces many different B cells. Each B cell type produces antibodies which react with a particular epitope. Note that the carboxyl ends of the heavy chain activate

B cells

complement and also binds to phagocytes. We will discuss this aspect of antibody function in a later chapter.

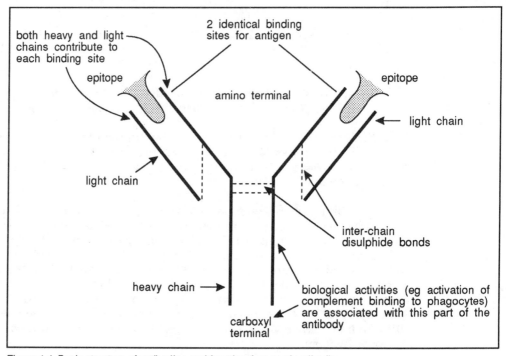

Figure 1.4 Basic structure of antibodies and functional parts of antibodies.

∏   Having examined Figure 1.4, make sure that you know which end of the antibody molecule binds antigen, which chains contribute to the binding of each epitope and how many epitopes can each antibody molecule bind and, finally, what part of the molecule possess biological activities such as being able to activate complement or bind to receptors on phagocytes.

1.4.3 Monoclonal and polyclonal antibodies

Let us examine the response to a protein molecule that possesses four non-identical epitopes A, B, C and D along its sequence (see Figure 1.5) which are recognised as being foreign or nonself by the host.

We mentioned earlier that it is the B lymphocytes that are responsible for producing these antibodies. Each B cell possesses many thousands of identical receptors that are essentially antibodies inserted into the membrane of each cell. Let us imagine that one B cell bears receptors which bind to the epitope A. Since epitopes B, C and D are different structures to A, this B cell only binds to A and we say that it is specific for A.

∏   Examine Figure 1.5. Before you read on, can you decide what are the major consequences of B cell activation by an antigen?

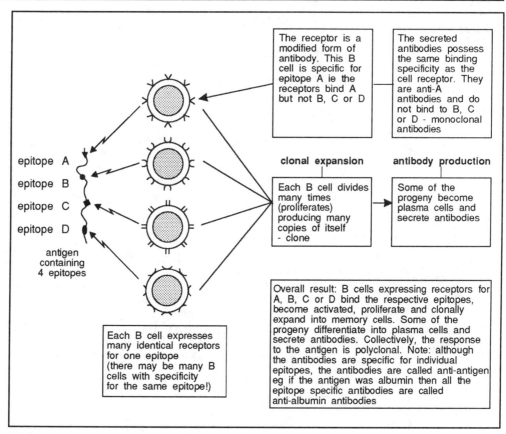

Figure 1.5 Antigens, epitopes and antibodies (see text for details).

Under the right conditions, the B cell specific for A will be activated and will undergo mitosis (proliferate) when it binds to the epitope. The resulting progeny of this cell will produce millions of antibody molecules all of which will have exactly the same specificity for A as the receptor on the parent B cell. The progeny are called a **clone** and the products **monoclonal antibodies**. The two consequences of B cell activation are therefore clonal expansion and antibody production. Other B cells may also possess receptors for the same epitope and will produce anti-A antibodies when stimulated. However, the binding sites of the other antibodies may be slightly different as the receptor may bind the epitope in a slightly different configuration. Additionally, some of these antibodies may only bind loosely to the epitope whereas others may bind tightly. We call these low affinity and high affinity antibodies respectively.

Other B cells will have receptors for epitopes B, C and D. So the response by the host to the whole antigen will be activation of many B cells of varying specificities resulting in the production of many clones producing many different antibodies to the four epitopes. We call this type of response a heterogeneous or **polyclonal** response and the mixture of antibodies **polyclonal antibodies**.

clone

monoclonal
antibodies

polyclonal
antibodies

| **SAQ 1.5** | Which one of the following statement is incorrect? |

Most environmental antigens

1) have many antigenic determinants (epitopes);

2) induce a polyclonal response;

3) can induce a polyclonal response to a single epitope;

4) induce both low and high affinity antibodies;

5) induce one type of monoclonal antibody per epitope.

1.4.4 The humoral response

humoral response

primary response

B lymphoblasts

plasma cells

The production of antibodies in an animal results in antibodies in the blood stream, hence the name humoral response (Figure 1.6). The first time the host is infected with antigen ie in the primary response, B cells specific for the many epitopes on the antigen(s) will be activated and undergo mitosis resulting in large dividing B cells called B lymphoblasts. Some of these cells will undergo an alteration in their morphology (differentiation) to become plasma cells which are packed with rough endoplasmic reticulum and are simply antibody factories releasing thousands of antibody molecules every second. These cells perform this task for a few days and then die.

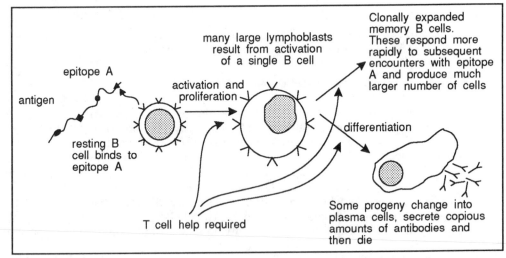

Figure 1.6 The basic humoral response to antigen. Note that T cells help the B cells respond. T cells are another group of white blood cells which we will discuss in later sections.

protective antibodies

Many of the antibodies remain functional in the body for a long time. We call these protective antibodies because they can quickly dispose of antigen if it invades the host a second time. Our milkmaids had protective antibodies produced in response to infection with cowpox which prevented development of smallpox.

Π See if you can write down a reason why the milkmaids were immune to smallpox virus because of their exposure to cowpox.

You may think, with some justification, that antibodies to cowpox would not protect an individual infected with the smallpox virus because the two viruses would have different sets of epitopes. We really do not know what virus Jenner used in his vaccinations but we can assume that there was some structural relationship between the two viruses ie they possessed some epitopes that were identical or bore a close similarity in structure. Hence the antibodies were able to bind either virus - the antibodies are said to be cross-reacting antibodies. The antibodies were able to intercept any extracellular smallpox virus in the milkmaids and the vaccinated individuals thus providing some protection against a full blown infection.

cross-reacting antibodies

Not only do individuals produce antigen specific antibodies during an immune response, they also expand the numbers of antigen specific B cells that originally respond to the antigen (Figure 1.6). All of the lymphoblasts we referred to do not become plasma cells; some of the progeny revert to small lymphocytes and become memory cells which remain in the body for long periods. These cells are ready to respond to a subsequent attempt by the antigen to invade the host. This increase in the numbers of antigen specific B cells is called clonal expansion.

memory cells

clonal expansion

You can see, then, that in a subsequent response (anamnestic response) the host will have many more antigen reactive B cells at its disposal and therefore will produce many more antibodies than in the primary response. The quality of the antibodies ie the overall affinity, also improves with repeated infections - a process known as maturation of the response thus improving the efficiency with which the antibodies deal with the antigen.

anamnestic response

maturation of the response

1.4.5 Antibodies enhance innate immunity

We said earlier (Section 1.3.5) that innate immunity is totally ineffective against many antigens. Since antibodies are invariably produced against most extracellular pathogens and many of these antibodies can then activate complement, antibodies compensate for these deficiencies and make most pathogens and other antigens targets for killing by complement or phagocytes which are attracted to the site.

Some antibodies provide an additional benefit in that they can opsonise the pathogen (in a similar manner to C3b's) for which they are specific and phagocytes expressing receptors for the non-antigen binding ends (carboxyl ends) of these antibodies then phagocytose the pathogen (Figure 1.7). It might be helpful for you to box in all of the innate immune components in Figure 1.7 to emphasis the relationship between antibody production and innate immunity. We deal in depth with complement and C3b in a later chapter.

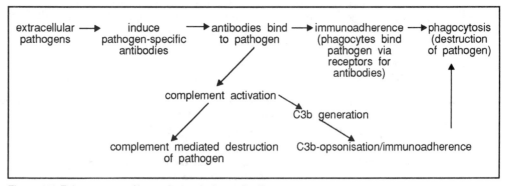

Figure 1.7 Enhancement of innate immunity by antibodies.

SAQ 1.6

Which one of the following statements is correct?

1) In a primary response, B cells once activated, clonally expand and become lymphoblasts which produce antibodies for a few days and then die.

2) Antibodies are bifunctional molecules in that there is an antigen binding carboxyl end and an amino terminal end involved in other biological functions such as activation of complement.

3) We can anticipate that practically all antibodies are to some extent cross-reactive.

4) Monoclonal antibodies are those produced in response to a single antigen, hence the name.

By now, you will appreciate that, whereas innate immune mechanisms alone are often ineffective due to the inability to recognise nonself, antibodies convert these into a potent force for destruction of extracellular antigens. We will see in Chapter 2 that antibodies possess many other characteristics that contribute to the protection of the host.

1.5 Acquired (adaptive) immunity - cell mediated immune mechanisms

1.5.1 Intracellular pathogens

The production of antibodies does not afford full protection to the host. As we have emphasised, antibodies deal very efficiently with extracellular antigens but some pathogens are rather invidious and hide themselves inside host cells where they cannot be attacked by antibodies or the various components of the innate immune system. These intracellular pathogens include all the viruses and some bacteria, fungi and protozoa.

1.5.2 T cells see cell-associated antigens

cell mediated immunity

T lymphocytes

We obviously require alternative mechanisms to combat these pathogens. These mechanisms are collectively called cell mediated immunity and the cell type primarily responsible for the majority of these activities are the T lymphocytes. Like B cells, T cells are antigen specific but they do not produce antibodies and they do not bind to epitopes on native antigens. T cells only recognise antigen as foreign when it is exposed on the surfaces of the host's cells attached to one of that unique set of molecules on an individual's cells we referred to earlier (1.2.1).

Major Histocompatibility Complex (MHC)

Human Leukocyte (Leucocyte) Antigens (HLA)

Let us briefly look at this class of molecules so that you can understand the basic function of T cells. The cell surface molecules are part of the Major Histocompatibility Complex (MHC) and they are called HLA (Human Leukocyte (Leucocyte) Antigens) molecules in Man and H-2 molecules in mouse. There are two classes of cell surface MHC molecules. In Man, each individual expresses many thousands of up to six different HLA Class I molecules on the surfaces of all nucleated cells and platelets and at least the same number of different HLA Class II molecules on a more restricted spectrum of cells.

Since all T cells recognise antigens only when associated with MHC molecules, we say that T cells are MHC restricted.

1.5.3 The killing of virally infected cells

cytotoxic T cells

Cells infected with viruses express viral antigens attached to MHC Class I molecules. A subpopulation of T cells called cytotoxic T cells (T_C) recognise this combination in an antigen specific manner and destroys the cell. This leaves the viruses without a host cell and exposes the virus to specific antibodies which may have also been generated during the response (Figure 1.8a).

At least some types of tumour cells also express tumour antigens (not recognised as self) associated with these MHC Class I molecules and this promotes killing by tumour antigen specific cytotoxic T cells. Foreign MHC molecules such as those expressed on cells of transplanted organs are also recognised by some cytotoxic T cells in the recipient and would be destroyed in the absence of immunosuppressive drugs. This is in addition to the destruction of allogeneic cells by antibodies in the recipient.

1.5.4 The activation of cells infected with bacteria, fungi and protozoa

cytokines
(lymphokines)

Cells with intracellular non-viral pathogens may express antigens derived from the pathogen along with MHC Class II (and in some instances MHC Class I) molecules. Another subpopulation of effector T cells (cell mediated immune cells - CMI cells), very similar in function to T helper cells (see below) release soluble messengers called cytokines (lymphokines) which bind to receptors on the infected cells. These promote various effects such as production of intracellular proteolytic enzymes which, in some cases, results in destruction of the pathogen. In this case, the infected cell is not destroyed (Figure 1.8b).

These are the two principle mechanisms by which cell mediated immunity operates; by either lysis of virally infected cells/tumour cells or by the activation of cells infected with non-viral pathogens. Notice, there is a direct recognition of self (MHC) as well as nonself (antigen). This contrasts to humoral immunity where there is only direct recognition of nonself (antigen).

T cells, on binding MHC + antigen, become activated and clonally expand in much the same way as in the B cell responses resulting in development of memory effector cells.

We would expect, then, that the milkmaids who were immune to smallpox would have developed memory cytotoxic T cells specific for MHC + cowpox antigen as well as specific antibodies. The immune T cells would make it difficult for the viruses to find a cellular home and the antibodies would intercept extracellular viruses. This combined protection gave them immunity to the virus.

1.5.5 T cell help and B cells

You may have noticed in Figure 1.6 that we included a requirement for T cell help in B cell responses. It is now accepted that the majority of B cell responses cannot proceed without T cell involvement. The antigen specific receptors on T helper cells, T_H, recognise antigen bound to MHC Class II on the antigen specific B cells and this interaction activates the T cell to produce cytokines (lymphokines) that support all stages of the B cell response including activation, proliferation and clonal expansion, differentiation and production of antibodies.

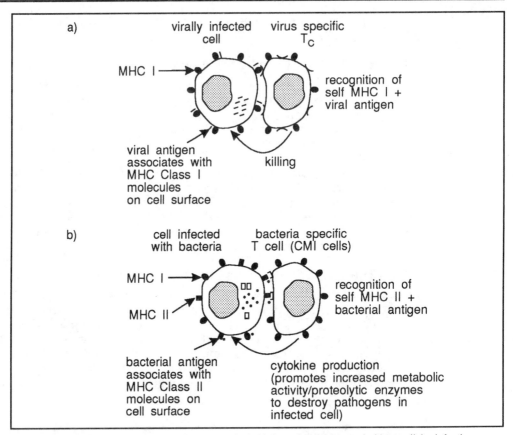

Figure 1.8 Cell mediated immune mechanisms. a) viral infected cell, b) non-viral intracellular infection.

SAQ 1.7

Complete the table by using + or - to indicate the presence or absence of each feature.

Feature	T$_H$	T$_C$
Express MHC Class I		
Recognises virally infected cells		
Produces antibodies		
Binds MHC II		
Kills viruses		
Binds MHC I on virally infected cells		
Binds cell associated antigens		

regulatory cells

effector cells

As well as T$_H$ cells which provide support for B cells responding to antigen, there is another subpopulation of T cells which act to turn off responses, the suppressor T cells or T$_S$. The T$_S$ and T$_H$ are called regulatory cells whereas the cytotoxic T cells and the cells providing help to other infected cells are effector cells.

Π In Figure 1.10 we have constructed a flow diagram depicting the interconnections between innate and acquired immunity. Before you look at this in Figure 1.10, try to construct your own diagram to show possible connections between innate immune mechanisms, humoral and cell mediated immunity. It will help you to build up a clear picture of the connections between the various systems.

1.6 Major characteristics of acquired immunity

You should by now be able to conclude what are the major distinguishing features of acquired immunity when compared to innate immunity. They are:

- specificity. This is the ability to discriminate between different antigens (epitopes);

- self: nonself discrimination. This is largely a property of the T cells. B cells, although capable of binding antigen, need T_H cells to fully respond and produce antibodies;

- memory. This is the ability to remember previous contacts with the antigen and to be able to respond more rapidly and more efficiently to a subsequent encounter due to the larger numbers of primed memory cells.

1.7 The learning of self

We know from the earlier sections that the cells and tissues of an individual are recognised as self. The question then arises as to how the immune system knows what is self.

1.7.1 Is recognition of self inherited or learned?

You can do a relatively straightforward experiment to answer this question. An adult female white mouse is pregnant. The foetal mouse is injected with human serum albumin (HSA) together with cells from a black mouse ie a different strain to the white. The MHC molecules on the cells of the two strains of mice are totally different and each would be treated as foreign by the other adult strain of mouse. The cells are called

allogeneic cells

neonate

allogeneic cells - cells from a genetically non-identical individual of the same species. The foetus develops and is born (we call the newborn a neonate). During its growth to adulthood we give regular injections of HSA.

Π Examine Figure 1.9a which depicts this experiment and try to decide why antibodies against human cells and not against allogeneic cells and HSA are produced by the adult mouse. Try also to give a reason for a single injection of allogeneic cells but multiple injections of HSA. We will give an explanation in the next section.

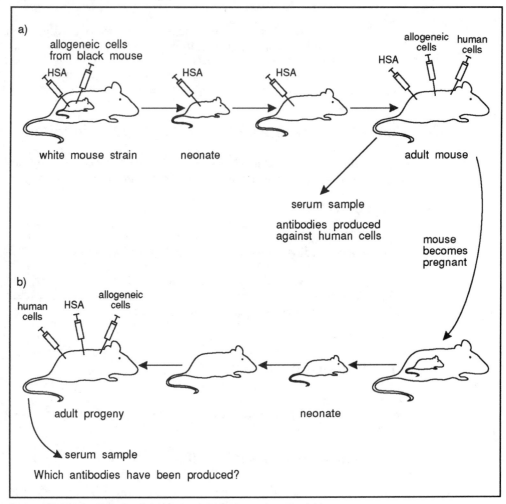

Figure 1.9 Recognition of self; learned or inherited - see intext activities.

1.7.2 Self tolerance

Did the fact that the adult mouse in experiment shown in 1.9a would only produce antibodies to the human cells surprise you? The reason is as follows. During foetal life the mouse has to learn what is self. Any antigen, which would normally be foreign to the adult mouse, if injected into the foetus, will be recognised as self along with the normal self cells and self proteins of the mouse. We say that the mouse has become tolerant or nonresponsive to these components and the antigens are called tolerogens.

tolerogens

However, there is one major rule of tolerance which must be adhered to if the mouse, when adult, persists in recognising these antigens ie allogeneic cells and HSA, as self. The rule is that tolerogens must persist throughout the life of the animal. We can use this to explain why we have to give multiple injections of HSA but only a single injection of allogeneic cells. HSA is a protein and would be treated like any other self protein in the adopted host. It would be continually replaced. But the recipient cannot synthesise HSA and so HSA would eventually be lost from the recipient. Thus, we need to repeatedly inject. HSA into the recipient to maintain HSA levels to ensure tolerance. With allogeneic cells the situation is quite different. Since they are living cells they continue to divide and respond in the adopted host and live quite happily alongside the cells indigenous to the white mouse. In other words, the allogeneic cells persist and will continue to be tolerised by the host and recognised as self.

We have however made an important simplification in the experiment described in Figure 1.9a. In practice this experiment might have a failure. If the allogeneic cell injection contained cytotoxic T cells these would have learned that the black mouse H-2 is self. When these cytotoxic T cells enounter the cells of the white mouse foetus, these cells would have different H-2 Class I components. The cytotoxic T cells would have recognised cells of the white mouse as nonself and would have attacked them. This is

graft-versus-
host reaction

called graft-versus-host reaction. If however the allogeneic cell injection did not contain mature cytotoxic T cells (for example if the cells were taken from the bone marrow of the black mouse) then there would be no graft-versus-host reaction. The bone marrow is a source of immature cells and contain T cells which have not yet learned to distinguish between self and nonself. If these are transferred to the white mouse, they would adopt the new environment and would accept the white mouse as self. As they mature in the white mouse, these allogeneic cells would no longer be able to recognise cells of the black mouse as self but would regard them as foreign.

Π Examine the experiment shown in Figure 1.9b and decide which antibodies the mouse would produce.

The answer is that antibodies would be produced against all three groups of antigens (human cells, HSA and allogeneic cells). This shows that the information on self recognition of HSA and alloegenic cells was not passed from mother to progeny.

We can conclude that recognition of self is not inherited but is a process that must be learned by each individual during development. Let us introduce you to a few rules of tolerance or self recognition:

• antigens introduced during foetal life or as a neonate are not recognised as foreign but induce a state of nonreactivity ie tolerance;

• this tolerance persists throughout life but only in the presence of the tolerogen;

• tolerance to self antigens is a learned process and is not inherited. It cannot be passed on to the progeny;

• since this is immune recognition of self, tolerance is dependent on a functional immune system;

• breakdown of tolerance leads to reactivity against self ie autoimmunity.

So we now have two definitions which you should know. Antigens which induce an immune response are immunogens; if they induce a state of nonreactivity they are tolerogens.

SAQ 1.8	Give brief answers to the following questions.

1) Why does a transplanted kidney continue to function for years in the recipient where it is foreign?

2) On a purely theoretical basis, how could you treat newborns so that they could accept a transplant from any unrelated donor when adult?

1.8 Epilogue on innate and acquired immunity

You have now had an introduction to basic immunological principles and the language of immunology. Make sure you have thoroughly understood the material in this chapter and the responses to the questions before you proceed any further. To help you we have provided a summary diagram of innate and acquired immunity (Figure 1.10) which summarises much of the information covered in this chapter. Use this figure to get an overview of the major aspects of immunity. Begin by reading Figure 1.10 from the box labelled immunity and follow the various branches of immunity. We will deal with each of these aspects in greater detail in subsequent chapters.

Before we look at these aspects, we will provide you with an overview of the anatomy and function of the major lymphoid tissue.

1.9 Anatomy and function of lymphoid tissues

1.9.1 Introduction

We have already concluded that for an effective immune response interactions have to take place between various cell types such as T and B cells and these have to come into contact with the antigen. Since the antigen invades the tissues at one entry point in the body and many of the T and B cells, especially memory cells are in the blood, there has to be some system which brings the cells and antigen together and this is the major role

primary lymphoid organs

of the lymphoid (or lymphatic) system. Some components of the lymphoid system are primarily involved in the production of fresh T and B cells and indeed all other blood cells and are called the primary or central lymphoid organs. These include the bone marrow and the thymus. The lymphoid organs which provide the locale for activation

secondary lymphoid organs

of T and B cells by antigen and the production of memory cells are called secondary or peripheral lymphoid organs and include the lymph nodes, spleen and other diffuse lymphoid tissue found in the walls of the gut, respiratory and urinogenitary tracts.

We shall now take a brief look at the production of various immune cells in the central lymph organs and then examine the relationship between blood and the lymphoid system and how the structural features of the peripheral lymphoid tissues facilitate responses of T and B cells.

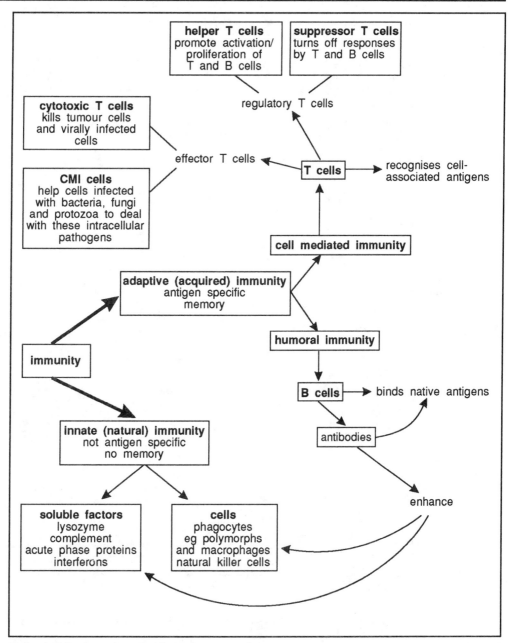

Figure 1.10 Innate and acquired immunity.

1.9.2 Bone marrow and haemopoiesis

stem cells

haematopoietic
cells

All the cells of the immune system originate from common ancestors called stem cells. These cells are self renewing and have the potential to become any type of cell found in blood - we say they are pluriopotential haemopoietic (or haematopoietic) cells (haema, haematoes means blood and poiesis means forming). Thus any one of these cells could become a T cell, a monocyte, a neutrophil or even platelets or red blood cells. During foetal life these cells first appear in the yolk sac and then in the foetal liver and finally in the bone marrow. These cells persist in the bone marrow developing into the various cell types throughout life.

lymphoid and myeloid pathways

The pathways for the development of the various cell types is indicated in Figure 1.11 and shows that the stem cell may choose the lymphoid pathway which results in the formation of T and B lymphocytes or the myeloid pathway which leads to development of red cells, platelets, granulocytes (such as neutrophils) and monocytes. There is also an undefined pathway leading to the production of Natural Killer (NK) cells. We shall take another look at haemopoiesis in Chapter 7.

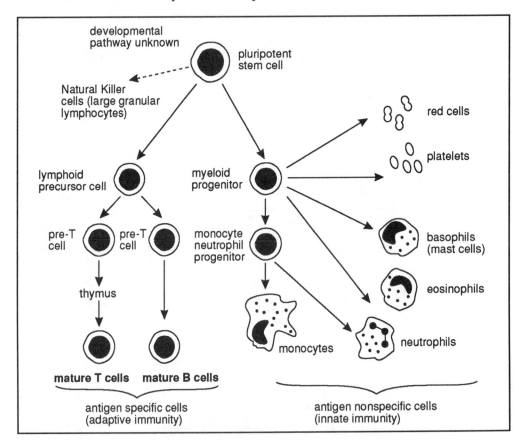

Figure 1.11 Origin of cells in the immune system.

1.9.3 Maturation of T and B cells and their migratory characteristics

T cells mature in the thymus

We think that B cells develop to maturity in the bone marrow in Man and these cells then move into the blood stream to be delivered to organs of the lymphatic system where they will remain until exposed to antigen. In direct contrast, T cells do not develop into mature competent cells in the bone marrow but must travel to another lymphoid organ, the thymus, to complete their development. Some of these cells live for many years and spend their time travelling between the blood, the tissues and the lymphoid organs looking for invading pathogens or antigens. Because of this continuous passage between blood and the lymphatic system they are called

recirculatory pool

recirculatory cells and form the major part of the recirculatory pool of lymphocytes. The remainder of the recirculatory pool is made up of memory B cells which result from activation of those B cells which originally migrated from the bone marrow to the various lymphoid organs. You should remember that of all the blood cell types only T and B cells ever return to the blood once they have left it. All other cells become either fixed tissue cells or are destroyed during inflammatory responses.

This recirculatory pool of antigen specific T and B cells represents a huge army of cells which can be targeted to any part of the body where antigen is localised. Some of these cells, as we saw earlier, will be specific for various epitopes on the antigen. As we have seen earlier in this chapter, this may result in production of antibodies which will mediate destruction of the antigen using complement and/or phagocytic cells or cell mediated immune mechanisms if the antigen is situated inside cells.

SAQ 1.9	Which one of the following is correct? 1) The lymphoid pathway is the development route for T and B lymphocytes and large granular lymphocytes. 2) The first multipotent stem cells can be detected in the foetal liver. 3) The recirculatory pool consists of T cells which continually travel between blood and lymph. 4) Memory B cells recirculate. 5) The bone marrow is a secondary lymphoid organ.

1.9.4 The monocyte series

Following development in the bone marrow, monocytes migrate into the blood stream and may stay in the blood for a few days before entering the tissues in response to localised antigen. On leaving the blood these cells are called macrophages and also professional phagocytes since their chief function is to phagocytose the offending antigens with the help of antibodies and complement. The macrophage is a very important cell in immunity and possesses other functions as well as phagocytosis. You will come across these cells throughout the text but particularly in Chapter 8.

macrophages (professional phagocytes)

fixed macrophages

Other monocytes migrate from the blood and become resident in various tissues; they are then called fixed macrophages. You came across examples of these earlier in the chapter. All the different macrophage cell types, whether mobile or fixed, are collectively referred to as the reticuloendothelial system. This plays a most important role in removing pathogens and antigens from the tissues and body fluids.

1.9.5 The granulocytes

The bone marrow maintains large stores of granulocytes so that they can be dispatched via the blood to localised antigen in tissues anywhere in the body at short notice. These cells, as the name suggests, contain granules or packets of enzymes or other mediators used in immune defence mechanisms. The neutrophils, so-called because they do not react to acidic or basic dyes and which form the greater part of the granulocyte pool, act in much the same way as macrophages and phagocytose antigen with the help of antibodies and complement. The antigens are then destroyed by hydrolytic enzymes.

Other granulocytes include the eosinophils and basophils which stain with acidic and basic dyes respectively. The eosinophils are particularly useful in defence against worms (helminths). The basophils (mast cells) are coated with particular antibodies and, on encountering antigen, degranulate releasing various pharmacological agents. Details of the properties and functions of these cells are given in Chapter 7.

1.9.6 Third population cells

killer and natural killer cells, lymphokine activated killer cells, antibody dependent cellular cytotoxicity

These cells include K (killer) cells and natural killer cells (NK) also called large granular lymphocytes (LGL). However, K cells have fallen out of favour and third population cells are now generally accepted to be NK cells or transformed NK cells (these are called lymphokine activated killer cells or LAK). Although the alternative name is LGL we cannot define the developmental pathway of these cells into either lymphoid or myeloid lines. NK cells have been found to kill some cells infected with virus and also some tumour cells but they are not considered to be antigen specific. Additionally, they can kill cells coated with certain antibodies, a process known as antibody dependent cellular cytotoxicity (ADCC).

∏   Now that you have accumulated quite a lot of information on the various cell types involved in immunity, organise this into a comparative table under the headings Cell Type, Antigen specificity? and Major Functions. Compare your table with Table 1.1 at the end of this chapter.

1.9.7 The peripheral lymphoid organs

You will, no doubt, have experienced 'swollen glands' situated under the jaw line when you have had a sore throat. Similarly, you may have noticed similar occurrences behind the knee (popliteal) with a lower leg abscess or infected cut. These lymph nodes are small encapsulated organs packed with cells, a large majority of which are T and B cells. During an infection more cells from your recirculatory pool enter these nodes and antigen specific cells multiply in response to the infectious agent which has been transported to the node. This results in enlargement of the node resulting in 'swollen glands'.

Lymphatics and transport of cells, proteins and antigens

These lymph nodes are situated in many parts of the body and are connected together by a network of vessels called lymphatics. These lymphatics originate from the interstitial spaces of the tissues of most parts of the body and are filled with interstitial fluid called lymph.

lymphatic system returns proteins to the blood

The lymph recovers many proteins lost from the blood in the vascular capillary beds and transports these proteins through the lymphatic system back to the blood. This is an essential function since proteins such as albumin cannot diffuse back into the blood against the protein concentration gradient in the capillary bed and must not remain in the tissues as this would draw fluid from the blood by osmosis resulting in accumulation of fluid or oedema. The amount of protein returned to the blood by the lymphatics in a single day represents up to one half of the total plasma proteins.

The lymphatics also act as a transport system for immune cells. The blood capillaries in these areas are extremely thinwalled being composed of only a layer of cells and a thin membrane. The T and B cells and indeed other cells such as monocytes and granulocytes can migrate across these walls into the tissues in response to a variety of chemical signals which we shall describe later. The T cells, monocytes and granulocytes may have been attracted to the site by the presence of antigen (pathogen) and they may induce an inflammatory response to get rid of it. Many of the T cells, B cells and monocytes finally make their way into the terminals of the lymphatics and are carried to a lymph node where the monocytes are retained many of them perhaps replacing fixed macrophages in the node. The T and B cells can continue along the lymphatic system back to blood for redistribution to other parts of the body.

The lymphatic absorbing terminals are, like the minute blood capillaries, composed of a layer of endothelial cells and membrane which are anchored to the tissues by delicate filaments. The lymph is moved up the lymphatics principally by muscle contraction which squeezes the vessels. In so doing it is thought that the endothelial cells separate forming small pores which allow the inlet of cells and proteins into the lymphatic vessel. Very much like the veins, valves are situated in the larger vessels to prevent backflow of the lymph.

SAQ 1.10

Match each of the items in the left column with one in the right column. Do not use any item more than once.

1) LGL	A) Phagocytic
2) Neutrophils	B) ADCC
3) Monocytes	C) Bone marrow cells
4) Memory cells	D) Antigen specific
5) Stem cells	E) Kupffer cells

Functions of lymph nodes

afferent lymphatics

Lymph nodes are encapsulated organs ie they are enclosed by a collagenous capsule. The lymph is delivered to the node from the interstitial tissues by afferent lymphatics and after flowing through the body of the node leaves by efferent lymphatics. Only about 10% of the T and B cells in the node arrive by the afferent lymphatics. Most T cells and B cells enter the node direct from minute blood capillaries supplying blood to the node. Many thousands of these cells pass through each node of the body every hour and are directed to defined areas of the node, which we will describe in Chapter 7. Here the lymphocytes can be activated by antigen which has also been carried from the tissues by the afferent lymphatics either free in the lymph, in monocytes or in specialised antigen presenting cells called Langerhans cells. We shall discuss the significance of these cells later in the text.

By making these huge numbers of cells pass through a node, the few cells expressing antigen specific receptors are able to make contact with the antigen and they become activated resulting in an immune response.

lymph node traps antigen

Most importantly, not only do the nodes provide the locale for interaction of T and B cells with antigen but it also is an effective filter which traps antigens in lymph thus preventing the antigens gaining access to the blood stream. These antigens are destroyed by cells of the reticuloendothelial system in the node. In other words, the nodes localise and destroy the antigen preventing its spread around the body. As indicated above, some antigen is taken up by Langerhans cells and some macrophages in the tissues and these cells then move into the afferent vessels and carry the antigens to the node. This cell associated antigen is used to stimulate an immune response.

Figure 1.12 illustrates the relationship between the blood capillaries, the tissues and the lymphatic vessels in the movement of cells, proteins and antigens into the lymph.

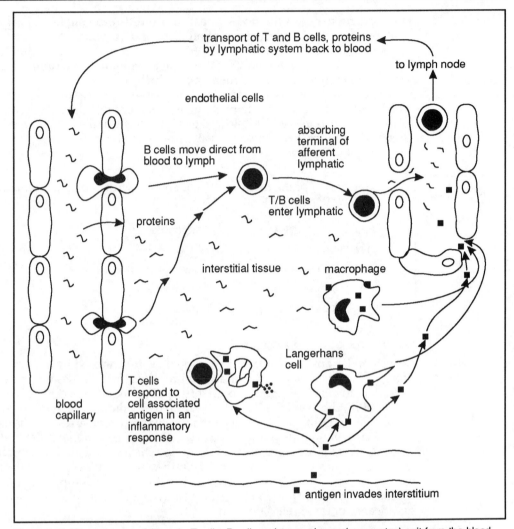

transport of T and B cells, proteins
by lymphatic system back to blood

to lymph node

endothelial cells

B cells move direct from
blood to lymph

absorbing
terminal of
afferent
lymphatic

T/B cells
enter lymphatic

proteins

interstitial tissue   macrophage

blood
capillary

T cells
respond to
cell associated
antigen in an
inflammatory
response

Langerhans
cell

antigen invades interstitium

Figure 1.12 Blood, tissues and lymph. T cells, B cells and macrophages (monocytes) exit from the blood by squeezing between endothelial cells. Most B cells and some T cells migrate across the interstitial tissues and enter the lymphatic for transport to the node. Some T cells may wander about the interstitium examining the various cell types for the presence of antigen. This may result in an imflammatory response. Some antigen may find its way to the lymphatic and be transported to the node whereas some will be taken by Langerhans cells and macrophages which will then move to the lymphatic. Proteins lost from the blood will also be recovered in the lymph. The T and B cells and the plasma proteins are carried up to the lymphatic system and returned to the blood.

thoracic duct

The lymph leaves the node via efferent lymphatics to other nodes further upstream. The lymphatics eventually fuse with other vessels to form larger vessels and these finally lead to the thoracic duct which pours its contents into the blood stream via the subclavian vein in the neck. T cells and memory B cells will finally leave the node they entered from the tissues or direct from the blood and will be carried along this system of nodes and vessels back to the blood. Antibodies, made in lymph nodes along the route, will also be carried to the blood to be distributed around the body.

| SAQ 1.11 | Which of the following statements apply to lymph nodes? |

1) They are supplied with lymph by efferent lymphatics.

2) They act as filters of the lymph by trapping antigen.

3) They are the sites for activation of T cells, all of which travel to the nodes via the afferent lymphatics.

4) They trap blood borne antigens.

Principal immune functions of the lymphatic system

In summary, the principal immunological functions of the lymphatic system are:

- it filters and traps antigens from body tissues thus denying access to other parts of the body;

- it provides the locale for T and B cells to make contact with antigen;

- it transports T and B cells from the tissues and lymph nodes back to the blood thus promoting recirculation of these cells.

The spleen and other peripheral lymphoid tissue

Just as the lymph nodes as filters across the lymph stream, the spleen acts in a similar manner to filter the blood stream and promote the interaction of antigen with immune cells.

There are also other sites in the body where we find less well organised lymphoid tissue. In contrast to lymph nodes and spleen which are contained with capsules, we find diffuse aggregates of immune cells called nodules or follicles in the walls of the gut beneath the epithelial cells. One example of these is the Peyer's patches in the small intestine. Similar aggregates are found in other mucosal tissues such as the walls of the respiratory tract. The tonsils represent another site containing aggregates of lymphoid tissue. Figure 1.13 gives you some example of the various types of lymphoid tissue distributed throughout the body.

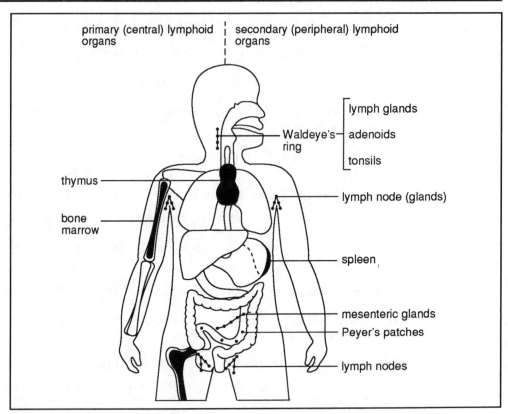

Figure 1.13 The major lymphatic organs.

1.10 Summary of the types of immune cells and their functions

To help you distinguish between the different cell types you have met in this chapter we have constructed Table 1.1 which names the cell types and defines their major functions. You will, of course learn much more about each cell type as you progress through the text.

Cell type	Antigen specific?	Major functions
T Lymphocyte		
Helper T cells (T_H)	Yes	promote responses by B cells resulting in antibody production and also activate other T cells
Cytotoxic T cells (T_C)	Yes	kill virally infected cells and tumour cells
Suppressor T cells (T_S)	Yes	suppress T_H and maybe also B cells
B lymphocyte	Yes	produce antigen specific antibodies which bind to antigen and promote its destruction
Third population cells		
Natural killer cells (large granular lymphocytes)	No	kill some virally infected cells and tumour cells. Also kill cells by antibody dependent cellular cytotoxic mechanisms
Myeloid cells		
Monocyte/macrophage	No	blood monocytes and mobile and fixed tissue macrophages phagocytose antigens especially those covered in C3b's or antibodies. Some macrophages also present antigens to T cells
Granulocytes	No	these include neutrophils which phagocytose antigen in a similar manner to macrophages and eosinophils and basophils which promote destruction of antigens by non-phagocytic means such as the production of cytotoxic factors and various pharmacological agents. Mast cells are the tissue equivalent of blood basophils.
Thrombocytes (platelets)	No	cells involved in blood clot formation and produce mediators similar to basophils/mast cells

Table 1.1 Properties of the major immune cells.

Summary and objectives

In this chapter we have provided you with an overview of innate and acquired immunity. You have learnt that the body can distinguish between self and nonself and that innate immunity is not antigen specific. We have explained that acquired immunity can be divided into two main types, humoral and cell mediated immunity. Humoral immunity acts through the production of antibodies produced by B cells. These antibodies may enhance innate immunity through a process called opsonisation. Cell mediated immunity depends on the activities of T cells. T cells can be divided into two types, effector cells and regulatory cells. The regulatory T cells (T helper and T suppressor) control the activation and proliferation of T and B cells. The effector T cells identify and kill tumour and virally infected cells (cytotoxic T cells) and may help to identify and stimulate the destruction of non viral intracellular infections (CMI cells). We have also provided you with an overview of the origin of the cells involved in immunity and with the layout of the lymph system.

Now you have completed this chapter you should be able to:

- explain what is meant by self and non-self and define epitope, antigen, immunogen and tolerogen;

- describe cellular and soluble innate immune mechanisms for eliminating antigens and explain the deficiencies of innate immunity;

- list and describe the major stages of the humoral response;

- discuss the differences between innate immunity and acquired immunity;

- describe the cell mediated immune mechanisms for dealing with intracellular pathogens;

- list the major features of tolerance.

- draw a diagram to illustrate the developmental pathways in haemopoiesis;

- list the major functions of the lymphatic system and describe the pathways B and T cells travel between blood, tissues and lymph.

Antibody structure and function

Antibody structure and function

In this chapter we will examine the structure and function of antibodies. We will begin by briefly examining the discovery of antibodies before exploring the structures of the various types of antibodies. We will then explore the biological functions of these important molecules.

This is quite a long chapter so do not attempt to read it all in one sitting.

2.1 Discovery of antibodies

Von Behring and Kitasato first described anti-toxins in 1890 for which they received the Nobel prize in 1902. This momentous discovery some 5 years after the first description of bacterial toxins by Roux and Yersin derived from a simple experiment demonstrating that sublethal doses of toxin injected into mice a few days prior to being given a lethal

immunisation

anti-toxins

dose fully protected the mice. They had been immunised (protected from) by the toxin resulting in the production of anti-toxins (antibodies) that blocked the lethal effects of the toxin. They also showed that serum transferred from these mice to non-immunised (naive) mice afforded similar protection to the recipient mice thus demonstrating that the serum contained the anti-toxins. This serum did not protect the recipient mice from

antibody
specificity

the lethal effects of unrelated toxins. We, therefore, say that the antibodies were specific for the toxin.

Π Examine Figure 2.1 carefully. It describes this experiment, the conclusions and defines some relevant terms. It would be a good form of revision to make yourself a list of the definitions and conclusions shown in this figure. For example active immunity - animal produces antibodies after being challenged by an antigen. Passive immunity - animal receives pre-formed antibodies from a donor.

passive
immunisation

This experiment also demonstrates the principle of passive immunisation. The transfer of serum from an immune individual to a naive one results in specific protection of the recipient from the effects of the toxin. In 1891, only one year later, this principle was clinically applied to a diphtheria patient passively immunised with horse antiserum containing anti-toxins to diphtheria toxin.

In 1900, Ehrlich called the specific serum component antikorper or antibody. In 1910, whole serum was salt fractionated and in 1939 Kabat and Tiselius electrophoretically separated serum proteins and found that most of the antibodies were in the gammaglobulin (γ-globulin) fraction as indicated in Figure 2.2. However, as you can see, some antibodies are detected in the beta (β) and (α) alpha regions so antibodies are

immunoglobulins

called immunoglobulins rather than gammaglobulins.

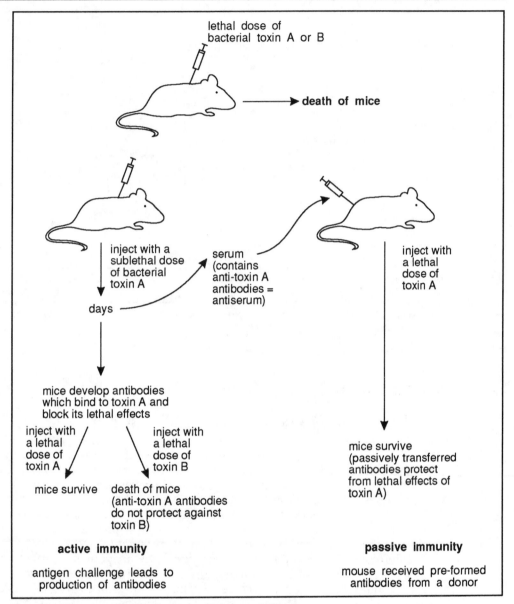

Figure 2.1 Discovery of antibodies by von Behring and Kitasato.

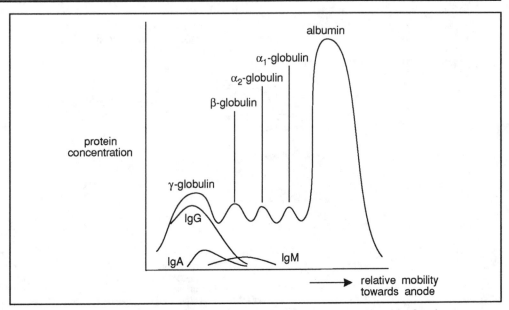

Figure 2.2 Electrophoresis of serum proteins showing the electrophoretic mobility of the 3 major serum immunoglobulins (IgA, IgG and IgM).

2.2 Antibody structure

2.2.1 Source of antibodies for structural studies

The antigen specificities and biological activities of the serum antibodies reflect differences in primary amino acid sequences of the antibody polypeptide chains and this heterogeneity presented the protein chemists with a major obstacle during studies of the structure of antibody molecules. However, electrophoretically homogeneous proteins found in serum of patients with plasma cell dycrasias, otherwise known as paraproteinemias or monoclonal gammopathies including multiple myelomas were found to be antibodies. These paraproteins or M proteins were, in fact, monoclonal antibodies produced by a single clone of malignant plasma cells and these products often comprise over 80% of the total serum antibody pool thus providing ideal material for structural studies. Additionally, some patients were found to pass subunits of antibodies into their urine. These proteins are the light chains, also called Bence Jones proteins and these were detected by heating the urine and inducing precipitation. These samples provided ideal material for studies on light chain structure. Similar plasma cell tumours in the mouse and rat provided material for the study of mouse and rat antibodies.

monoclonal
gammopathies

Bence Jones
proteins

SAQ 2.1

Which one of the following is correct.

1) Antibodies are generally negatively charged molecules.

2) Each antigen induces the production of antibodies of a single specificity.

3) Bence Jones proteins are whole antibody molecules otherwise known as M proteins.

4) IgG is the most common antibody in serum.

5) Transfusion of blood from a donor who is of blood group A to a recipient who is also blood group A does not induce antibodies in the recipient. (Note group A refers to certain antigens on the surface of red blood cells. The other groups are B and O).

2.2.2 The four chain monomeric unit of antibody structure

The basic structural components of antibodies were identified before any of the monoclonal antibodies described in 2.2.1 were available. However, it was found that treatment of serum with 40% ammonium sulphate would precipitate the antibody fraction from whole serum. Porter and others, therefore, immunised rabbits with bovine serum albumin (BSA) producing anti-BSA antiserum and obtained a crude fraction of anti-BSA antibodies by ammonium sulphate fractionation followed by further purification using ion exchange chromatography. The following experiments on this material led to the 4 chain model of antibody structure.

Determination of molecular weight

Ultracentrifugation studies on the antibody fraction gave a sedimentation coefficient of 7S that corresponds to a molecular weight of about 150 000 Daltons (150 000 D).

Effects of reducing agents

The antibodies were subjected to mild treatment with the reducing agent 2-mercaptoethanol which cleaves exposed disulphide bridges (S-S) reducing the cystines to cysteine residues. The exposed sulphydryl groups were then alkylated using iodoacetic acid to prevent the antibody reforming and the resulting polypeptides were separated by gel filtration in the presence of acetic acid to prevent hydrophobic interactions ie association between polypeptide chains. This method is based on separation of proteins by molecular weight. Two polypeptides were eluted from the column one with a molecular weight of 50 000 D and one with a molecular weight of 25 000 Daltons and were called heavy (H) and light (L) chains respectively. They were present in equimolar proportions thus indicating a 1:1 ratio in the intact molecule. Based on the molecular weight of the whole antibody, the data showed it consisted of 2 H and 2 L chains ie the basic structural unit of antibody was H_2L_2.

heavy chains

light chains

Proteolysis of antibody molecules

papain
proteolysis

Early studies using the enzymes papain and pepsin provided essential information on the structure and function of the antibody molecule. Papain proteolysis at neutral pH resulted in 2 identical fragments called Fab (Fragment - antigen binding) and a single Fc piece (Fragment - crystallisable). Both types of fragment were of about 50 000 Daltons molecular weight and the yield of Fab was about twice that of Fc. The Fc piece is further degraded with prolonged exposure to papain.

Fab possessed antigen binding properties whereas Fc did not. The Fab fraction, however, could not precipitate antigen suggesting only a single antigen binding site (we will examine this aspect in a little more detail a little later). The Fc crystallised out of solution in neutral pH, hence the name. Antiserum generated against Fab reacted with both heavy and light chains whereas anti-Fc could only bind heavy chains.

pepsin digestion

F(ab')$_2$

pFc' fragment

When the antibody was subjected to pepsin digestion in acid pH two major fragments were generated: the F(ab')$_2$ which included the two Fab pieces joined together by a disulphide bond and a pFc' fragment. The F(ab')$_2$ was found to bind two epitopes exhibiting bivalency and would precipitate antigen. This is because the fragment could link up different molecules of antigen into a large complex which would not remain in solution.

∏ Examine Figure 2.3 carefully. It summarises the information given in the text and see if you can explain why F(ab')$_2$ but not Fab can precipitate the antigen.

The answer is that F(ab')$_2$ has two binding sites and can therefore cross-link antigens to form an insoluble network. Thus:

Fab has only a single binding sites and cannot form such a network.

You should be able to see from Figure 2.3 that antibody molecules are made up of a number of repeat sequences - note the same loop structures along all of the chains.

∏ Which step in Figure 2.3 indicates that the interchain S-S holding the two Fab's together in F(ab')$_2$ is weaker than the interchain S-S joining the heavy and light chains?

You should have concluded that the treatment of F(ab')$_2$ with mild reducing conditions leads to the production of Fab as a result of reducing the S-S bonds holding the two Fab's together. Under the same conditions the S-S bonds holding the heavy and light chains together are not reduced.

Based on this evidence the four chain structure $H_2 L_2$ was accepted as the basic structural unit of all antibody molecules. This evidence is summarised in Figure 2.3 and also points out other antibody fragments derived by proteolysis.

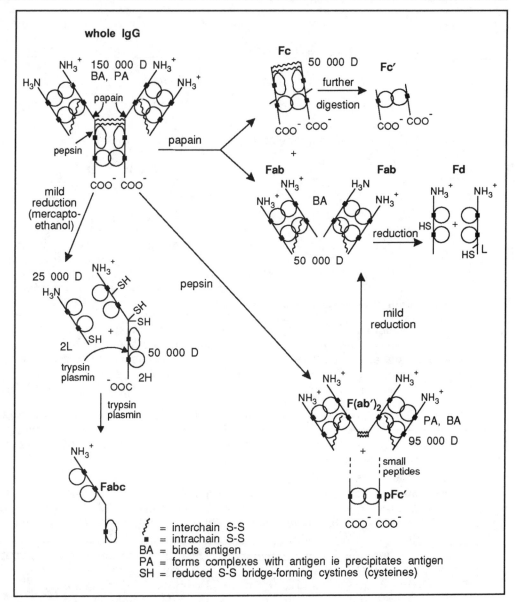

Figure 2.3 Reduction and proteolysis of antibodies.

Indicate which of the following fragments would precipitate antigen.

1) Fab.

2) IgG treated with 2-mercaptoethanol.

3) Fc.

4) IgG treated with pepsin.

5) Reduced and plasmin-digested antibody.

2.2.3 The domain structure of antibodies

Use the stylised drawing of an antibody shown in Figure 2.4 to follow the description of antibody structure given below. There are some surprising features when one examines an antibody heavy chain. The first thing that strikes you is that the chain appears to be divided into 4 repeat units of about 100-110 amino acids in length each of which contains an intrachain disulphide bridge that encloses a loop of about 60-65 amino acids. Similar repeat units, two in number, are seen in light chains. Each of these units are folded into a globular arrangement called a domain.

The domains at the amino terminal end of the light and heavy chains are characterised by a great degree of sequence variability and are referred to as variable domains being designated V_L and V_H respectively. These domains are responsible for the antigen binding function of the molecule, there being two binding sites each composed of one V_L and one V_H domain.

The remainder of the light and heavy chain domains are relatively constant, that of the light chain being designated C_L, the 3 constant domains of the heavy chains C_H1, C_H2 and C_H3. The amino acid content of the heavy chain constant domains is remarkably similar amongst different antibodies. Some 30% of the positions have the identical amino acid in place.

Situated between C_H1 and C_H2 domains is the non-globular hinge region high in proline and possessing varying numbers of disulphide bonds between the heavy chains. Flexibility at this point allows a certain amount of independent movement of the antigen binding sites.

The individual domains interact with their opposite numbers on the adjacent antibody chain through non-covalent interactions and this makes a major contribution to the tertiary structure and conformational stability of the intact immunoglobulin molecule. Thus V_L and V_H domains interact forming the antigen binding site; C_H1 and C_L closely associate, as do the two C_H3's of the heavy chains. Notice that there is no association between the two C_H2 domains. The exception to this is in IgM and IgE where the non-association occurs in the C_H3 domains.

margin notes: domain / variable domains / constant domains / hinge region / non-covalent interactions / antigen binding site

Π Having carefully examined Figure 2.4, close your text and construct a diagram of an antibody molecule similar to Figure 2.3 which more clearly shows the disulphide loops. Label the following features: light and heavy chains and their molecular weights, intrachain and interchain S-S, V_H and V_L, C_H1-3, the antigen binding site, Fab and Fc, hinge region. Make sure you indicate the association between domains excluding C_H2's. Indicate the sites for proteolytic cleavage by pepsin and papain and which end of the molecule confers biological activity such as complement activation. When you have completed your diagram, check it against Figures 2.3 and 2.4.

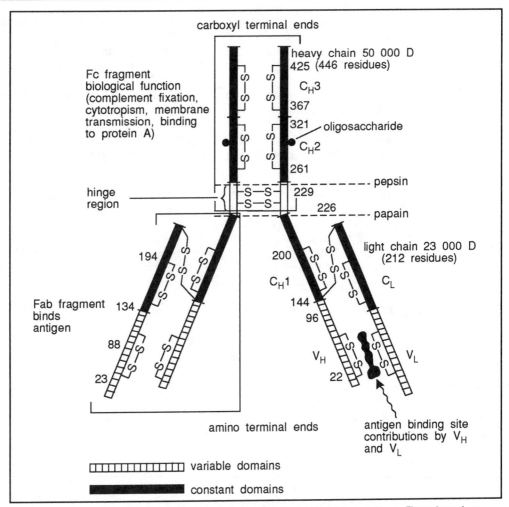

Figure 2.4 Stylised chain structure of human IgG showing disulphide bridge positions. Figure based on human myeloma protein Eu. Note that in the text we have confined ourselves to discussing the protein part of an antibody. Most antibodies are however glycosylated and have short oligosaccharides attached to their heavy chains. These increase the solubility of these molecules.

2.2.4 Antibody heavy chain isotypes

5 antibody classes

isotypes

Due to major differences in the amino acid sequences of the constant domains of the heavy chains, based on serologic and chemical studies, it has been possible to identify 5 antibody classes in most mammals IgG, IgM, IgA, IgD and IgE possessing gamma (γ), mu (μ), alpha (α), delta (δ) and epsilon (ϵ) heavy chains respectively. These major differences in the heavy chains are called isotypes. In other words, the gamma chain is one isotype and the alpha chain is another and so on. Antibodies to these heavy chains ie anti-isotypic antibodies, are often called anti-globulins in the literature.

∏ Work out a protocol for producing antibodies to, say, the gamma heavy chain. You will come across the answer in Figure 2.6 and its associated text.

These heavy chains vary in molecular weight from 55-77 000 Daltons, the μ and ϵ chains possessing 4 constant domains rather than the 3 C_H domains of the γ and α chains. The δ chain, although having 3 constant domains, possesses an intermediate molecular weight due to an extended region between C_H1 and C_H2. Note we can use a more specific nomenclature $C_δ1$ is the first constant domain in a δ chain; $C_γ2$ is the second constant domain in a γ chain. Typically domains consist of about 110 amino acids.

As we shall see later, there are further subdivisions of some of the classes into subclasses based on gross amino acid differences. However, these subclasses are more closely related to each other within an Ig class than to other classes. IgG, IgE and IgD each consist of the basic 4 chain monomer whereas IgM is a macroglobulin consisting of 5 monomers linked together. IgA exists in various forms but principally occurs as a dimer.

anti-isotype
antibodies

All individuals of a given species possess identical heavy and light chains (however, see allotypes, idiotypes below). Although there is homology between human IgG and, say, rabbit IgG there are large differences in the amino acid sequences of the heavy and light chains of the two species. These differences are recognised by the rabbit as being 'nonself' and hence antigenic. Therefore immunisation of the rabbit with any of the human antibodies or the heavy chains would result in rabbit antibodies directed to the human antibodies ie anti-isotype antibodies.

∏   List the possible uses of anti-isotypic antibodies. Check with Figure 2.5.

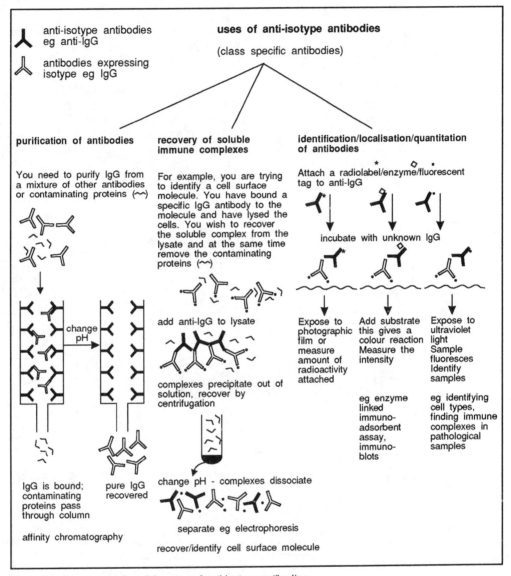

Figure 2.5 Some examples of the uses of anti-isotype antibodies.

2.2.5 Light chains

light chains

The light chains exist in two distinct forms, lambda (λ) and kappa (κ) isotypes. These are further divided into subtypes. These light chains are common to all the antibody classes; however, each antibody molecule possesses either two λ chains or two κ chains, never a mixture, as a single B cell cannot produce both. The two light chain isotypes are significantly different; only about 35% of the two sequences exhibit homology (ie are the same). In contrast, human and mouse kappa chains are about 60% homologous with each other.

The proportion of λ to κ chains in antibody molecules varies from species to species. In man 70% of light chains are κ; in mouse 95%. In many species such as sheep, goat, cow, horse and chicken over 95% are of the λ type.

Π   Practically, you would have a problem if you immunised the rabbit (Section 2.2.4) with whole IgG to produce antibodies specific for the gamma chain only. Can you think why? Check your answer with Figure 2.6.

You should have noted that even though we immune the rabbit with pure human IgG, a variety of antibodies are produced. Some recognise the epitopes on the L chain and these antibodies will bind with human IgA, IgM, IgD and IgE since all human antibody classes carry the same L chains. To obtain antibodies from the mixture which are specific for IgG, we have to remove the antibodies which bind with L chain epitopes. This is done by affinity chromatography using a column containing L chains or by adding excess L chains to block the anti-L antibodies (see bottom part of Figure 2.6).

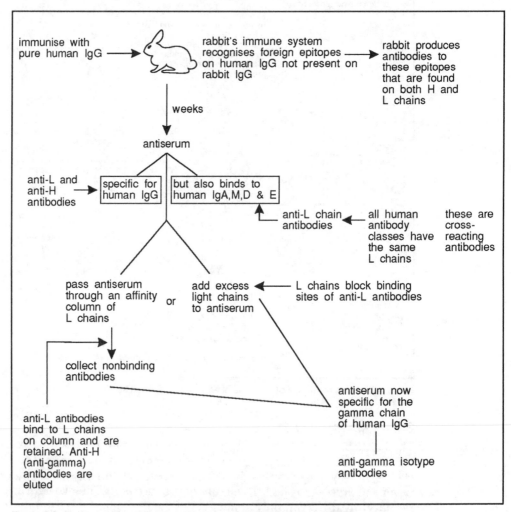

Figure 2.6 A scheme for producing anti-human isotype (gamma) antibodies.

2.2.6 The heavy-light (H-L) chain disulphide linkages

inter H-L
disulphide
bonds

We have already referred to the principal types of disulphide bonding in antibody molecules. Special mention needs to be made about those inter-H-L S-S bonds as there are some unique arrangements in some Ig classes. The C-terminal cystine in kappa chains and the penultimate (last but one) cystine in lambda chains invariably link with the neighbouring heavy chain. The point of attachment on the heavy chains in the other antibody classes appears to vary considerably. In human and murine IgG1 the linkage is at residue 220 after the second domain. In all the other subclasses of IgG as well as in IgM, IgE, IgD and IgA1, the attachment point is at the junction between the variable and first constant domains, about 80 residues closer to the amino-terminal end of the molecule (see Figure 2.4).

SAQ 2.3

Which one of the following statements is correct?

1) The molecular weight of the IgM is about 900 000 D.

2) IgG, IgA and IgE each consist of the 4 chain monomeric unit, IgD is a dimer and IgM a pentamer.

3) Variable domains are found on both heavy and light chains and are situated next to the hinge region.

4) The IgG H-L disulphide link is attached to the junction between V_H and C_H1.

2.2.7 Bifunctional antibodies

The structural studies just described should help you to appreciate the relative ease by which the antibody molecule can be broken up into individual chains by reduction and the susceptibility of the nonglobular portions such as the hinge region and sequences between the domains to proteolysis while maintaining the integrity of the domains.

It is now thought that these domains evolved to subserve a particular function. Investigations on the resulting fragments have shown that as well as the variable domains expressing antigen binding properties, the constant domains possess a variety of biological activities that, for example, promote the inflammatory response resulting in the destruction of the invading antigen and placental transfer of some antibodies to protect the infant. These functions are thought to be mainly localised in the second and third constant domains.

We can say, then, that antibodies are bifunctional; the amino terminal end or F(ab')$_2$ possessing an antigen binding function and the carboxyl terminal end or Fc expressing a variety of biological functions. In many cases, the binding of antigen is a prerequisite to the expression of the biological function. To remind you, again refer to Figure 2.4.

SAQ 2.4

Match each item from the left hand column with one of the features in the right column. Do not use any item from the right hand column twice.

1) Hinge region	A) Isotype
2) Antigen combining site	B) 110 amino acids
3) Domain	C) Precipitates antigen
4) IgM	D) (V_H V_L)
5) F (ab')$_2$	E) High proline content
6) Heavy chain	F) Polymer

2.2.8 Special structural features of antibody classes

Π We are now going to provide quite a lot of information concerning the structure of the various groups of antibodies. Although we provide drawings of some of these, it would be a useful exercise to draw these for yourself using the information provided in the descriptions. In this way, you will get a much better idea of these structures than you would from merely passively looking at our figures.

IgG

four subclasses of human IgG

In normal human adults IgG comprises about 75% of the total serum antibodies. Only about 50% of the total body IgG is intravascular; the remainder is distributed around the body in extracellular fluids. There are four subclasses of human IgG (IgG1, IgG2, IgG3 and IgG4) of which IgG1 comprises two thirds and IgG2 one quarter of the total IgG. The molecular weight of molecules of this class is 146 000 D except for IgG3 which is 170 000 D due to an extended hinge region. IgG is low in carbohydrate, with only a single oligosaccharide attached to each $C_\gamma 2$ domain. Remember our convention of nomenclature, $C_\gamma 2$ is the second constant domain in a γ chain).

The sequence differences between the four subclasses is much less than that between the major classes; in fact there is about a 90% homology in the constant domains of all subclasses.

The number and nature of inter-chain disulphide bonds in the IgG molecule is a discriminating feature among the subclasses. The unique positioning of the L-H linkage in IgG1 has already been referred to. The number of interheavy disulphide bridges in the hinge region is also variable, numbering two in IgG1 and IgG4, four in IgG2 and up to 15 in IgG3.

Π These features are summarised in Figure 2.7. There is a lot of information on this figure some of which will not make much sense to you at this stage. The figure has been provided as a summary and you will be able to use it for revision. At this stage particularly notice the long half lives of most of the IgG subclasses. See if you can write down a reason why this is important in the transfer of antibodies from mother to foetus across the placenta (placental transfer).

As the mother is exposed to pathogens and other antigenic materials, she will produce antibodies against them. Many of these will be IgGs. To be able to protect the foetus from these pathogens, the antibodies need to be able to stay in the foetal circulation for a long time. Only IgG antibodies are resistant to proteolysis. They are therefore suitable for this task. We will learn of some additional reasons a little later.

IgM

IgM constitutes about 10% of total serum immunoglobulins and about 80% of all IgM is intravascular. The vast majority of IgM molecules are in a pentameric form of the basic 4 chain monomer with a molecular weight of about 900 000 although small amounts of monomeric IgM are present in normal serum and in higher amounts in certain pathologic sera.

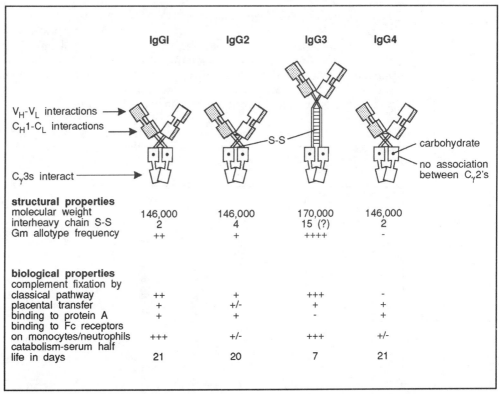

Figure 2.7 Structure and properties of IgG subclasses. Note that we have summarised some important biological properties of these antibodies. Note Gm = gamma marker. We will deal with these biological properties in detail later.

additional constant domain

The subunits of IgM possess an additional constant domain C_H4 or $C_\mu4$ and there is also an additional tail piece of 18 amino acid residues at the carboxyl end of each heavy chain. This extra domain is thought to represent a gene insertion of a domain at the C_H2 position. Hence the second and third constant domains in IgG are homologous to the third and fourth in IgM.

There is no defined hinge region and no inter-heavy chain disulphide bridges between the first two constant domains. In Man, these are found at the end of the $C_\mu2$ domain (residue 337) and in the penultimate position of the $C_\mu4$ domain (residue 575). The monomers are joined together by disulphide bridges in the $C_\mu3$ domain (residue 414) in Man and at the penultimate cystine position (residue 575) in mouse.

J (joining) chain

An additional peptide chain, the J (joining) chain is found disulphide-linked between two heavy chains of two separate monomeric units using the penultimate cysteines (residue 575). This cysteine rich 137 amino acid long peptide is attached to the pentameric structure in the IgM-producing B lymphocyte and is though to regulate the polymerisation process prior to secretion.

About 12% of IgM is carbohydrate. There are five oligosaccharides attached to each heavy chain by asparagine residues distributed among the 4 constant domains. These are thought to prevent interactions between the monomeric units.

Electron micrographs of pentameric IgM depicts the molecule as having a dense central portion representing the carboxyl terminal domains with radiating arms of F(ab')₂ units. Although there is no hinge region in IgM, the molecule is flexible probably between the second and third constant domain and can even assume a 'crab-like' appearance when binding to multiple epitopes on, say, a bacterial surface. Although IgM is pentameric and possesses potentially 10 antigen binding sites, IgM is functionally pentavalent using one binding site per subunit.

The major structural features of the IgM subunit are shown in Figure 2.8. To help you to get a feeling for the size of antibody molecules we have included some dimensions in this diagram.

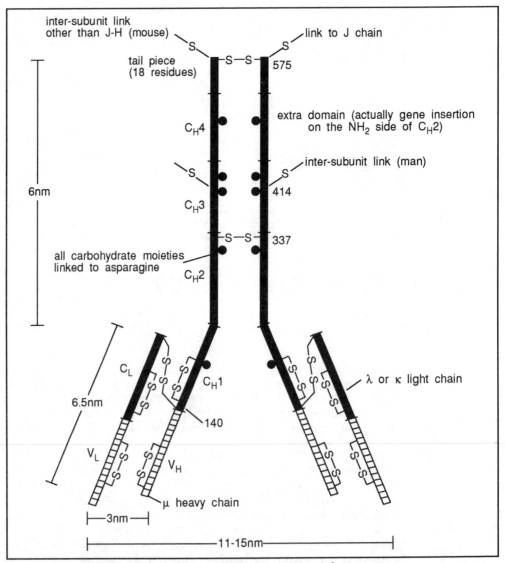

Figure 2.8 Basic structure of monomeric IgM. Sizes in nonametres (10^{-9}m). Numbers refer to amino acid residues from the N terminal.

<table>
<tr><td>

SAQ 2.5

</td><td>

Indicate with a + each of the properties from the left column that describe IgG and/or IgM.

Property	IgG	IgM
1) 4 domains/heavy chain		
2) Intersubunit S-S		
3) Tail piece		
4) Kappa chains		
5) J chain		
6) Pentavalent		
7) Interheavy chain S-S		

</td></tr>
</table>

IgA

IgA is the second most commons serum antibody after IgG. It can exist as a monomer or as polymers of 2 and 3 basic subunits. It possesses three constant domains as in IgG but, in common with IgM, possesses a carboxyl terminal octadecapeptide containing a penultimate cysteine. Dimeric IgA is the predominant immunoglobulin found in external body secretions and as a polymer possesses a single J chain disulphide linked to the penultimate cysteines of one heavy chain of each of the two subunits. There are unique additional intrachain disulphide bonds in the first two constant domains of IgA and also cysteines of unknown function in the $C_\alpha 2$ domain.

external body secretions

Secretory IgA (sIgA) has a molecular weight of about 380 000 Daltons and is made up of 2 four-chain units joined by one J chain. An additional peptide, the secretory component (SC) is also present. SC is a single chain glycoprotein with a molecular weight of about 70 000 D and has a total of 20 cysteine residues and 7 oligosaccharides all of which are linked to asparagine. It is not known how SC is attached to the IgA dimer. The favoured model is one in which SC is joined by disulphide bridges to the two alpha chains of one subunit only.

secretory IgA

secretory component

B plasma cells in the submucosa secrete IgA held in the dimeric form by J chain. This is then bound by secretory component (SC) acting as a receptor for IgA on the submucosal surface of epithelial cells. SC then facilitates the transport of sIgA across the epithelial cell and out onto the mucosal surface where it provides a first line of defence against invasion by pathogens or other antigens that threaten to invade by crossing the epithelial layer.

mucosal surface

There are two major subclasses of IgA, IgA1 and IgA2, in Man, goats, monkeys and rabbits and one in rats and mice. IgA1 is predominant in serum whereas IgA2 is the principal IgA in secretions. There are two subtypes of IgA2. One found in Caucasians, IgA2 (A2m(1)) is unique in having no H-L S-S bonds; the other IgA2 (A2m(2)) is found in Mongoloids and Negroids and has conventional interchain disulphide bonds. IgA1 had been found to be susceptible to proteolysis in the hinge region by extracellular enzymes produced by species of *Neiserria, Hemophilus* and *Streptococcus* which probably explains why sIgA2 is predominant in secretions as it is resistant to proteolysis.

H-L S-S bonds

IgE

IgE is the so-called reaginic antibody present in normal serum at extremely low levels. It exits as a monomer of molecular weight 190 000 Daltons, there being 4 constant domains like IgM. Other IgM-like properties include the absence of a hinge region, the

reaginic antibody

non-association of the two $C_\epsilon 3$ domains and the degree of glycosylation. The heavy chains are linked by 2 disulphide bridges on either side of the $C_\epsilon 2$ domains.

IgD

IgD is present in trace amounts in serum. Its H chain is structurally similar to IgG in having a V_H, a hinge region with one interheavy chain disulphide bridge and 3 constant domains. By contrast, IgD contains high amounts of carbohydrate distributed in many oligosaccharide units and possesses unique octapeptide tailpieces at the carboxyl terminal ends of the heavy chains.

octapeptide
tailpieces

A mystery surrounds the structure of murine (mouse) IgD since it was discovered that the whole $C_\delta 2$ domain is missing in this species. The hinge is only about half the length of human IgD and there does not appear to be any H-H disulphide bonds in this region.

SAQ 2.6

Complete the comparative table below.

Characteristic	IgG	IgM	sIgA	IgD	IgE
1) No of subclassess		1			
2) No of domains/molecule	12				
3) Presence of J chain (+/-)	-				
4) Heavy chain (Greek letter)	γ				
5) Molecular weight (kD)	146			180	
6) Extended COO- end? (+/-)	-				
7) Monomers/molecule	1				
8) κ or λ chains/molecule	2				

2.2.9 Allotypes

As well as the isotypic variations resulting in antibody classes referred to above, increased structural heterogeneity is provided by genetic variants of the antibody constant domains.

These normally represent short sequence differences amongst individuals of the species and are the products of alternative forms (alleles) at a given gene locus that are inherited in Mendelian fashion. Therefore, not all members of the species possess any given allotype. In Man, these allotypes have been found in IgG (called Gm allotype or gamma marker) and IgA (Am allotype or alpha marker) heavy chains and kappa (Km or kappa marker) light chains only. As an example, individuals with Gm(4) allotype have arginine at residue position 214 on IgG1 whereas Gm(4)⁻ individuals do not.

Another example of a Gm (Gamma marker) on IgG1 is G1m(1) that denotes the sequence Asp-Glu-Leu at residues 356 - 358. The sequence for those individuals not possessing this marker is Glu-Glu-Met. These minor amino acid differences can be immunologically detected by individuals not possessing the particular allotype.

Π Is it true that immunising an individual with material antibodies would not result in the production of anti-isotype antibodies? (Justify your answer).

It is true. Isotypes reflect the gross structure of heavy chains of each antibody class which are the same in all members of the species. Antibodies to isotypes can only be generated in another species. But it is also true that anti-allotype antibodies might be produced if the material antibodies are of different allotype to the recipient. A mother can become sensitised to paternal allotypes on fetal antibodies during pregnancy and she will produce anti-allotypic antibodies. Rheumatoid factor (often IgM), present in the sera of some rheumatoid arthritis patients, binds to IgG from sera of selected normal individuals indicating the presence of an allotype.

2.2.10 The variable region and antigen binding

Examination of a wide spectrum of antibodies of different specificities and myeloma products from patients showed many differences (heterogeneity) in the amino acid sequences of the first 110 amino acids of light and heavy chains, namely the variable regions.

If you plot changes of amino acids at each residue over the variable sequences of many different antibodies, the resulting variability plot would look something like that shown for heavy chains in Figure 2.9. A similar pattern emerges for the light chain. You will notice that most of the variability is not random but appears to be confined to three regions of the variable sequence in the proximity of residues 30-35, 50-65 and 95-100 although this varies from antibody to antibody. These regions have been called Hypervariability Regions or, more recently, Complementarity Determining Regions (CDR) due to their involvement in binding to the antigenic epitope situated within the binding site of the antibody. The 4 intervening sequences are called Framework Regions (FR). These are relatively invariant and contribute to the overall 3-dimensional structure of the combining site. Since there are 3 CDR's in both light and heavy chains, the antigen combining site could involve up to six major interactions with parts of the epitope.

Complementarity Determining Regions (CDR)

Framework Regions (FR)

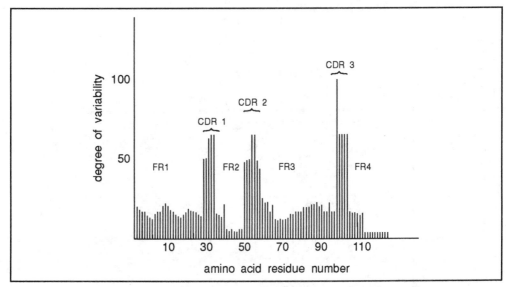

Figure 2.9 Variability plot of amino acids in V_H. FR = Framework Regions; CDR = Complementarity Determining Regions (see text for details).

So what do we know about the relative contributions of the H and L chain to the binding of antigen? Well, separated L chains appear to retain little antigen binding capacity whereas isolated H chains do retain some affinity for the antigen, albeit much less than the whole antibody from which they were derived. If the isolated H and L chains are allowed to reassociate at physiological pH then most of the capacity of the original antibodies is regained. It would seem therefore that H chains contribute the major part of the antigen binding function of the Fab site.

The next question is how do these six CDR's come together to form a binding site? In the past few years, X-ray diffraction studies have shown that all domains appear to have a common folding pattern with several straight segments of polypeptide running antiparallel to each other along the axis of each domain. These segments are kept separate by hydrophobic amino acid sidechains. There are 9 such segments in variable domains and 7 in constant domains. The folds or loops at the ends of some of the segments carry the CDR's so these are brought close together. There is a close association of the segments of the H and L chain variable regions thus causing close apposition of all 6 CDR's.

Π A very simple analogy is for you to hold a ball between the tips of the middle 3 fingers of both hands. One hand represents a heavy chain, the other a light chain and the fingers represent the folding of the 2 variable domains. If you can imagine polypeptide segments running up the outside of each finger to the tips, with antiparallel segments running away from the tips along the inside surface of each finger. The sharp bend at the tip represents the loops carrying the CDR's, all six making contact with the epitope (the ball).

Now look at Figure 2.10. It is a rather stylised representation of a folded variable domain and the possible interactions of the CDR's from both light and heavy chains with the antigenic epitope. These are non-covalent interactions involving hydrogen bonding, ionic and hydrophobic interactions. The size of the binding site has been estimated using ligand competition and electron spin resonance studies to be able to accommodate at most a linear hexasaccharide sequence.

non-covalent
interactions

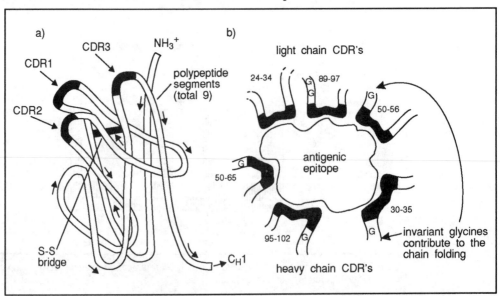

Figure 2.10 a) V_H folding of polypeptide segments showing apposition of CDRs. b) Interactions of CDRs from H and L chains with antigenic epitope.

2.2.11 Idiotypes

To complete the nomenclature for immunoglobulin structure we have to mention idiotypes. This is a rather fancy name for describing short sequences, mostly in the CDR's but sometimes outside them, present in an antibody with a defined specificity. If you immunise one animal with specific antibodies (call these Ab1) raised against an antigen epitope in a genetically identical animal, the antiserum of the recipient will contain antibodies (Ab2) specific for the binding site of the donor (Ab1) antibodies. Ab2 will bind to sequences within the V regions of Ab1 related to the antigen specificity. These sequences are collectively called idiotypes.

idiotypes

anti-idiotype antibodies

In other words, the variable region, in particular the CDR's of Ab1 are acting as an antigen inducing the production of Ab2 antibodies. Ab2 antibodies are known as anti-idiotype antibodies. Each short sequence recognised as antigenic is called an idiotope. All the idiotopes make up the idiotype. That part of the idiotype to which the anti-idiotype antibody binds is called the paratope. The anti-idiotype antiserum may contain some antibodies directed to idiotopes inside the binding site and the binding of these antibodies will be inhibited by antigen. Other antibodies may be directed to idiotypes outside the binding site and will not be inhibited by antigen (see Figure 2.11).

paratope

idiotopes

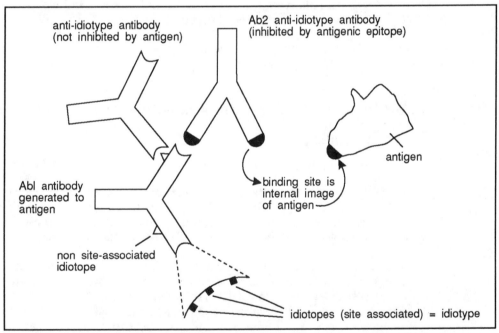

Figure 2.11 Idiotypes and anti-idiotypes. Note that the anti-idiotype antibody may compete for the binding site with the antigen. We have indicated two types of idiotopes. One set is not associated with the binding site of the antibody (represented by Δ), the other set (represented by ■) is associated with the binding site (see text for further discussion).

Anti-idiotype vaccines?

As is shown in Figure 2.11 the binding sites of some of these Ab2 antibodies will closely resemble the epitope for which Ab1 was specific. We say that the anti-idiotype is the internal image of the antigen. You perhaps could extend this idea and suggest that you could use Ab2 to immunise an individual against the antigen. The idea is that by innoculating with Ab2, this will stimulate the production of antibodies which will bind the antigen. This has been done. Additionally, in some cases where the antigen has been

internal image

a hormone, Ab2 was able to bind to the hormone receptor and mediate similar effects as would hormone binding. This is obviously a promising area for anti-idiotype vaccines.

Idiotypes are often unique to the products of a single B cell clone (private idiotype) or may be shared with other clones (public idiotypes). Idiotypes are, in some cases, inheritable.

SAQ 2.7

Choose which of the following statements is/are true or false.

1) Allotypes represent differences in the variable regions principally of IgG which are inherited.

2) You can generate anti-idiotype antibodies to human idiotypes allogeneically (in other individuals) or xenogeneically (in other species).

3) Heterogeneity in antibodies is due equally to allotypic, isotypic and idiotypic differences.

4) Anti-allotype antibodies are found in some individuals who have never been exposed to the allotype.

5) Idiotypes are only expressed on some isotypes.

2.3 Interaction of antibodies with antigen

2.3.1 Specificity

The manner by which a particular antibody molecule is able to discriminate between one antigenic epitope and another closely related epitope is called the specificity of the antibody. Ideally, the antibody will bind best to the antigen that elicited its production but all antibodies also bind to closely related epitopes albeit with less efficiency. However, the same epitope or a closely related epitope may be present on two different antigenic molecules; if the antibody binds to both this is an example of cross reactivity.

Karl Landsteiner was the first to study specificity of antibodies. He and his colleagues immunised rabbits with horse serum proteins to which he had attached the ligand meta-azobenzene sulphonate. The rabbit antiserum was tested for its reactivity to the same ligand or closely related ligands attached to chicken serum proteins. The idea of using chicken serum proteins in the test was to restrict the reaction to only those antibodies that were specific for the ligand. You should realise that antibodies would have also been generated to other epitopes on the serum proteins as well!

∏   What alternative method might Landsteiner have used to eliminate the effects of anti-horse serum protein antibodies from the experiment? Examine Figure 2.12 carefully and you should find the answer.

Figure 2.12 is quite complex and contains a lot of information. If you start at the top you will see that we have represented the meta-azobenzene sulphonate attached to the horse serum protein. We have used the words hapten and carrier which need explanation.

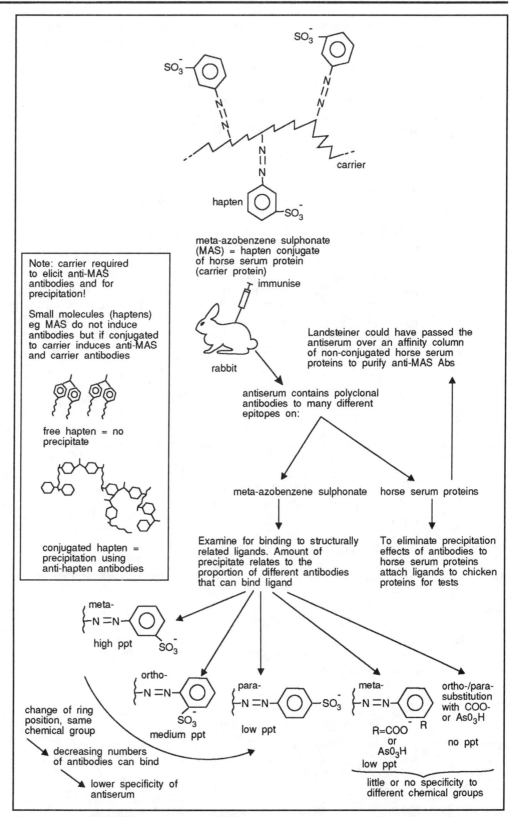

Figure 2.12 Landsteiner's experiments on specificity.

A hapten is a structure which is by itself unable to stimulate the production of antibodies but if associated with another structure (a carrier) is able to do so. Thus in the case of the experiment illustrated here, meta-azobenzene sulphonate acts as a hapten. Notice that the antibodies that are produced in response to injection by the meta-azobenzene sulphonate: horse serum protein conjugate shows some cross reactivity to similar components conjugated with serum proteins. Notice that Landsteiner could have eliminated the effects of anti-horse serum protein antibodies from the experiment by passing the antiserum through a column containing horse serum proteins (affinity chromatography). The anti-horse serum protein antibodies would have been removed from the antiserum.

Thus Landsteiner's showed that most reactivity was observed when the antibody was tested against the ligand used to elicit the antibodies but there was some cross reactivity with structurally related groups especially when they were attached to the meta position. In other words he demonstrated that the antibodies possessed a substantial amount of specificity for the immunising ligand.

Reconsider what comprises the antigen-binding site on the antibody molecule. There are 6 CDR's each capable of interacting with a complementary part of the antigen. If all six strongly interact, then the antibody could be said to be highly specific for that particular epitope. Small changes either in conformation or chemical grouping within the epitope may result in only 3 CDR's interacting with the epitope. The antibody will be less specific for this epitope; since it still binds this is an example of cross reactivity.

2.3.2 Affinity defined

The measure of strength of binding of antibody with an epitope is called the affinity. Obviously, the involvement of all 6 CDR's in binding the epitope will result in a high affinity. Thus represents the sum total of all the non-covalent interactions between a single antibody combining site and the epitope and can be measured.

Methods for the measurement of binding affinity involve allowing antigen and antibody to come to equilibrium in solution and then measuring the concentrations of the two reactants [Ab] and [Ag] and the product [Ab-Ag]. The interaction can be represented by:

$$[Ab] + [Ag] \rightleftarrows [AbAg]$$

According to the Law of Mass Action, the association constant is given:

$$K = \frac{[Ab-Ag]}{[Ab][Ag]}$$

Now, when half the antibody combining sites of the antibodies are bound by antigen, [Ab] = [Ab-Ag] and

$$K = \frac{1}{[Ag]}$$

affinity constant   For K, (the affinity constant), to be large, the concentration of antigen is small. In other words, high affinity antibody only needs low concentrations of antigen to achieve filling of half of its binding sites.

2.3.3 Avidity

As we have said, affinity is really a measure of the binding strength of a single antigen binding site. However, antibodies are multivalent ie IgG has 2 binding sites, IgM has 10 and many antigens, especially micro-organisms may express many repeating units of the same determinant. This provides the opportunity for the antibody molecule to use more than one binding site and obviously this increases the strength of attachment to the antigen since for dissociation to take place all binding sites must let go of the antigen at the same time. The potential total binding strength of an antibody is termed avidity. This is dependent on the affinities of each binding site but surprisingly, is not the sum the affinities. In other words, the avidity of IgG is not twice the affinity of one binding site. Indeed it has been estimated that bivalent binding of antigen by IgG may represent a thousand fold increase in binding strength over that of IgG utilising only one binding site. It would appear therefore that interaction with antigen at one binding site influences the affinity of the second binding site.

IgM is of special interest here as it is generally accepted that the affinities of individual binding sites is low. However, since IgM has many valancies, this is offset by multivalent binding and IgM is an extremely avid molecule.

SAQ 2.8

Fill in the spaces in the following statements using the words and phrases provided below. (Note some words may be used more than once).

1) The antigen combining site is composed of one [   ] domain and one [   ] domain. Each domain contributes 3 [   ] that interact with an [   ] on the antigen. The 3-dimensional configuration of the site is maintained by four [   ] in each domain.

2) Avidity is a measure of the binding strength of the [   ] antibody molecule and is estimated to be much higher than the [   ] of a single antigen binding site.

3) The [   ] domains of an antibody can be [   ] , resulting in the production of an anti- [   ] antibody. Each [   ] may consist of several [   ] and the site recognised by the second antibody is called a [   ] .

Word and phrase list

idiotype, idiotopes, paratope, whole, variable, complementarity determining regions, V_H, V_L, affinity, framework regions, epitope, idiotype.

2.4 Biological functions of antibodies

So let us examine the consequences of antibody binding to antigen. Take, for example, what happens when a bacterial species infects an individual. B cells recognise epitopes on the bacterium as foreign and secrete antibody molecules. These antibodies can then attach to the micro-organism and promote its destruction. These two functions are attributable to the Fab and Fc ends respectively and we shall now deal with how these activities fulfil a protective role in the host.

As you are reading this section try to appreciate how well designed antibody molecules are and how diversification into isotypes provides protective roles for different sites of the body.

2.4.1 Direct consequences of antigen binding

Π Even if all the antibody did was to bind to the antigen, how would this fulfil a protective role for the infected individual? Before you read on list the possible consequences.

The principal consequence is that the antibody may block some function of the antigen or pathogen. Two antibody classes are specially fitted for this type of role, IgG and IgA.

IgG

anti-toxin   IgG is particularly efficient as an anti-toxin due to its ability to capture soluble toxin molecules by high affinity binding sites. Most toxins mediate their effects inside cells and enter through receptors on the host cell. Binding of IgG prevents this. As we shall see later, this also promotes the destruction of the toxin because of another unique property of IgG.

sIgA

Due to its strategic location on the mucosal surfaces of gut, respiratory tract, urine tract, and in saliva, tears, vaginal secretions and prostatic fluid, sIgA makes contact with pathogens before they gain entry to the body.

sIgA is able to bind invading bacteria inhibiting their attachment, thus preventing their colonisation. Virus specific IgA may prevent virus entry into epithelial cells by blocking virus-receptor interactions. sIgA also is thought to bind many harmful substances in the gut preventing their passage across the mucosal membrane. The sIgA coated pathogens eg (viruses, bacteria) or protein antigens are then passed to the exterior by defaecation from the gut. Similarly the harmful entities can be removed from the respiratory tract or by mucociliary action.

Notice that these activities do not result in the destruction of the antigen but simply prevent their entry into the host. sIgA does not activate complement (Section 2.4.3) and is therefore unlikely to promote any cytotoxicity against bacterial pathogens. We emphasise that the actions described here are protective roles involving only the Fab portion of the antibody and not killing of the pathogen.

This functional expression of antigen binding appears to be the principal activity of sIgA with little known Fc mediated biological functions. The major role of the Fc of sIgA, then, may be in providing a site of attachment for the secretory component thus facilitating passage of sIgA to the external surfaces to perform this protective function.

| **SAQ 2.9** | A patient has a history of persistent respiratory infections. Tests reveal that his B cells respond normally to antigens and his T cell counts are normal. His blood group is AB and he has normal levels of serum IgG and IgA. These symptoms best fit with which one of the following: |

1) The patient is not producing antitoxins.

2) An inherited deficiency to produce IgM.

3) Low serum levels of specific IgE.

4) Failure to produce a secretory component.

5) Non-production of IgA.

2.4.2 Fc mediated complement-independent functions

If the pathogen manages to evade the sIgA mechanism and invades the host then the antibody has to promote killing or destruction of the antigen to eliminate it from the host environment. Notice the word 'promote' since antibody alone cannot kill any pathogen!

This activity is mediated by the Fc and there are two pathways by which this can be carried out. The first is receptor-mediated involving cellular receptors for Fc of some antibody classes, the second involves Fc activation of a complex series of serum proteins collectively called complement.

Fc receptor-mediated phagocytosis

Fc receptors

Some cells possess surface receptors for the Fc's of IgG and IgE in man. There is also some evidence for Fc receptors for IgA on neutrophils. There are three types of Fc receptors for human IgG1 and IgG3 called $Fc_\gamma RI$, $Fc_\gamma RII$ and $Fc_\gamma RIII$. The latter is a low affinity receptor for the gamma chain and is also called CD16. $Fc_\gamma RI$ is found on macrophages. $Fc_\alpha RII$ is found on macrophages, neutrophils and eosinophils. $Fc_\gamma RIII$ is found on macrophages, neutrophils, eosinophils and mast cells.

opsonisation

Apart from mast cells, the other cells use these Fc receptors for phagocytosing antigen. As indicated above, IgG1 or IgG3 specifically binds to pathogens, a process known as opsonisation or forms complexes with soluble antigens. The exposed Fc's then become bound by the Fc receptors on the phagocyte and the pathogen or immune complex is engulfed by the cell by invagination of the cell membrane to form a phagosome. This fuses with an enzyme-containing lysosome and the contents of the resulting phagolysosome are digested resulting in destruction of the antigen.

Fc mediated cell lysis

antibody dependent cellular toxicity

When some cells or pathogens are coated with specific antibodies particularly of the IgG and IgE classes, they may be lysed rather than phagocytosed by a number of different types of cells. For instance, liver fluke larvae coated with antibodies are lysed by eosinophils that engage the Fc of the coating antibodies via specific receptors. Antibody-coated measles virus-infected host cells are lysed by macrophages in a similar manner. Cells of transplanted tissues may be killed in this manner by macrophages, neutrophils or another cell type called the NK (natural killer) cell, also known as the large granular lymphocyte. This process is known as antibody-dependent cellular cytotoxicity.

Fc receptors for IgE

Mast cells and basophils (Chapter 7) express high affinity receptors for IgE. This type of antibody, produced locally in the submucosa and skin, attaches to Fc receptors on mast cells. On binding antigen, the IgE induces the mast cell to degranulate resulting in release of histamine, leukotrienes and chemotactic factors. These cells also possess receptors for complement components and the consequences of release of these mediators is discussed below.

2.4.3 Fc-mediated complement activation

Complement is composed of 11 proteins found in inactive forms in serum and interstitial fluid and is produced on site by cells such as macrophages. Six components comprise the classical pathway of activation induced by the presence of antigen complexed with antibodies of the IgG and IgM isotype only. There is an alternative pathway of complement activation that is independent of antigen-antibody complexes. Either of these pathways can then proceed on to the membrane attack complex resulting in lysis of bacteria or other cells. Figure 2.13 is a complex figure so use the descriptions given in the text to help you follow it. Begin with the antibodies shown in the bottom left hand corner. Note that complement proteins are identified by C.

The complement cascade

In the activation and amplification steps of the classical pathway the components are present as inactive proenzymes that acquire proteolytic activity when they themselves have been cleaved by an enzyme activated in the previous step. This sequential activation is also called the complement cascade. In contrast, the formation of the membrane attack complex does not involve proteolytic steps.

The classical pathway

The classical pathway is initiated by complement component C1q being bound by 2 Fc regions of 2 adjacent IgG (except IgG4) molecules or one IgM molecule. The concentration of IgG molecules on the surface of, say, a micro-organism has to be high to ensure that 2 molecules are close enough together to bind C1q. Since a single IgM is made up of the equivalent of 5 IgG molecules, you can see that IgM is much more efficient in activating complement than IgG.

The binding of C1q to the antigen-antibody complex then activates another component of C1 called C1r which then cleaves C1s. The cleavage product is a protease that can cleave 2 components, first C4 and then C2. The major fragment resulting from C4 proteolysis is C4b and this attaches to the surface of the micro-organism. C2 cleavage produces a large fragment C2b that hydrophobically attaches to C4b. The resulting C4b2b is called C3 convertase that cleaves C3 into two major fragments C3a and C3b.

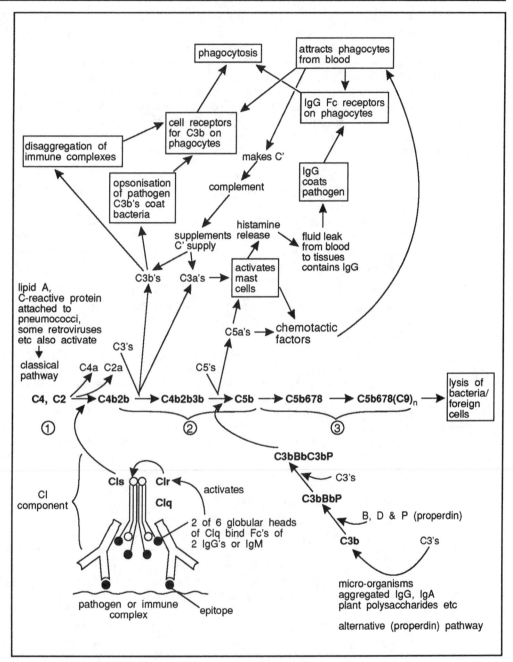

Figure 2.13 Complement activation and the biological consequences (see text for details).

anaphylatoxin

Both these fragments play major biological roles in inflammation. C3a is called an anaphylatoxin since it binds to mast cells (Chapter 7) inducing their degranulation resulting in release of many mediators but most importantly, histamine and neutrophil chemotactic factor. Both mediators diffuse towards the nearest venule. The histamine induces withdrawal of the venule epithelial cells from each other resulting in the opening of small pores in the blood vessel wall through which fluids escapes from the blood into the tissue. This may contain antigen specific IgG and more complement which will amplify the reaction towards the tissue located antigen. The neutrophil chemotactic factor attracts neutrophils to exit from the blood and move towards the source of the factor ie the antigen-antibody complex.

neutrophil
chemotactic
factor

The many C3b's, generated during the cleavage of C3, fall on to the surface of the antigen. If this happens to be a bacterium, the cell is coated (opsonised). Phagocytes such as neutrophils are attracted to the site (see above). Some possess receptors CR1 for C3b and are able to bind the C3b coated antigen resulting in phagocytosis. This attachment of particles via C3b is called immune adherence (immunoadherence). A similar mechanism operates for antigen-antibody complexes. In addition, C3b's tend to break up large complexes making them more suitable for phagocytosis.

immune
adherence

amplification
step

A small number of C3b's also attach to C4b2b forming C4b2b3b - a C5 convertase. This is the end of the so called amplification step of complement activation.

The alternative pathway

This is initiated by a variety of activators such as complexes of IgG, IgA and IgE, various micro-organisms, some virus infected cells and a variety of chemicals and is summarised in Figure 2.13. C3 is the most abundant component in serum and is being constantly converted to C3b. If this can be stabilised on surfaces of, say, micro-organisms, it binds a serum protein called serum protein B that is then cleaved by another protease called protease D and this results in a complex C3bBb. This is stabilised by another factor, called factor P (properdin) to give C3bBbP that is analogous to C4b2b. This then cleaves another C3 to produce a C5 convertase C3bBbC3bP analogous to C4bC2bC3b of the classical pathway.

properdin

Membrane attack complex

Both complexes C4b2b3b and C3bBb3bP are able to cleave the next component of complement C5 to C5b and C5a. The latter is another anaphylatoxin similar to C3a but a much more powerful agent. In addition to binding to mast cells, it is a chemotactic factor for neutrophils and also activates neutrophils and produces an increased expression of C3b receptors. Hence, C5a not only attracts neutrophils to the site but promotes immune adherence and phagocytosis.

The C5b molecule attaches to the surface of the micro-organism and promotes the attachment of C6, C7 and C8. This complex C5b678 somehow influences polymerisation of a number of C9 molecules to form a bore hole through the membrane of the micro-organism, osmotic pressure is lost and lysis of the cell occurs.

Π   This is quite a complex series of interactions. You might find it useful to re-read Section 2.4.3 and this time draw out your own scheme then check that you have drawn a sequence similar to ours shown in Figure 2.13.

Π Examine Figure 2.13 carefully and notice how complement uses a variety of pathways to focus on the destruction of the pathogen. Work out how many different pathways exist in the system.

SAQ 2.10

Select which of the following are true or false.

1) C3a is an anaphylatoxin that activates mast cells and is also chemotactic for neutrophils.

2) Opsonisation only involves the coating of antigens by C3b particles.

3) Similar levels of IgG and IgM activate complement.

4) Only immune complexes activate the classical pathway.

5) The membrane attack complex is dependent on the generation of C5b.

2.4.4 Other Fc related functions

Catabolism of antibodies

protease
resistance

It is noticeable that only IgG appears to be resistant to protease action in body fluids that result in destruction of antibodies. This resistance lies in the Fc region. Most subclasses of human IgG have a half life of about 20 days which means that IgG antibodies produced in response to infection persist for a long period. In contrast, the other antibody classes are rapidly degraded. Because of this persistence, IgG can be

protective
antibody

considered to be the protective antibody and most successful vaccination procedures result in high levels of high affinity IgG antibodies that will remain active in the individual for a very long time.

Placental transfer

IgG antibodies are unique in being able to cross the placental membrane into the foetal circulation; this is mediated by a sequence in the $C_\gamma 2$ or $C_\gamma 3$ domains. These antibodies will provide protection against some common pathogens for the child during foetal and neonatal (newborn) life during the period when the child's immune system is not fully operational. Notice the evolutionary selection of antibodies that are resistant to proteolysis for this task.

You will remember that sIgA, namely IgA2, appears to have been selected for its resistance to proteolysis by micro-organisms (Section 2.2.8). This is present in high quantities in mother's milk and protects the infant against common gut pathogens during the early months of life. As we have said (Section 2.4.1) this function can be ascribed to Fc as it is the attachment of the secretory component to Fc that it responsible for its presence in external body fluids including milk.

Π Make three lists of the biological functions of antibodies. In the first, list the functions which are independent of Fc. In the second, list the functions which depend upon Fc but are independent of antigen binding. In the third, list the functions which depend both upon Fc and on antigen binding. Then use Figure 2.14 to check your response. This figure summarises the many biological functions of the antibody classes.

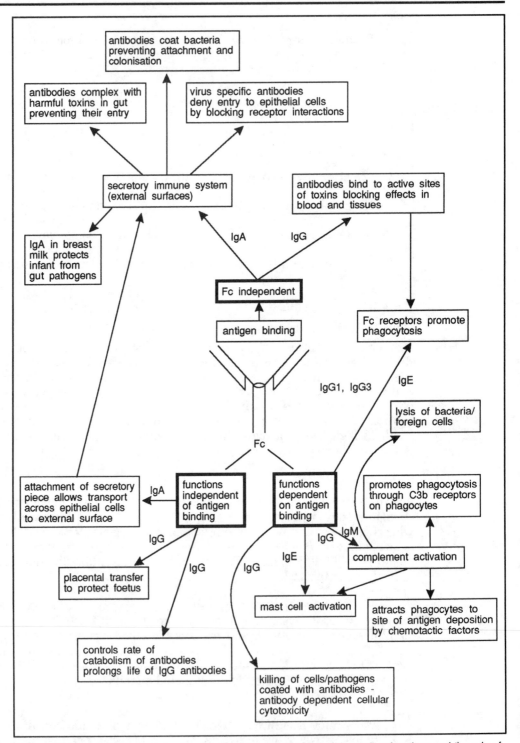

Figure 2.14 Bifunctional nature of antibodies. The Fc independent and dependent functions and the role of the antibody classes (isotypes).

SAQ 2.11

Which of the items in the left column would block any of the four activities? Indicate by a + those functions you would expect to be still fully intact use +/- if significantly reduced and - if completely absent.

Factor	Fc mediated phagocytosis	Lysis of bacteria	Chemotactic activity	C3b mediated phagocytosis
1) IgA-antigen complexes				
2) Properdin deficiency				
3) IgG deficiency				
4) Mast cell malfunction				
5) C4 deficiency				
6) C5 deficiency				

2.5 Antibodies as cell membrane antigen receptors

The body's population of B cells express many thousands of receptors for antigens and these are a form of antibody of the same specificity as the secreted antibody product. Each B cell therefore produces receptors for a single epitope. Newly formed (virgin) mature B cells express on their surface two receptor types IgM and IgD. Only monomeric antibodies act as receptors ie you never see the pentameric form of IgM or the dimeric form of IgA on a B cell. There are minor structural differences between membrane and secreted antibody, the principal one being the presence of an elongated carboxyl end on the heavy chains to accommodate anchorage to the membrane. More about this in Chapter 3.

2.6 Antibodies as biotechnological tools

You now know quite a lot about the structure and function of antibodies. Antibodies have many practical applications, for example in therapeutics and diagnostics. This involves many changes to antibody molecules. These are discussed in detail in the BIOTOL text 'Technological Applications of Immunochemicals'. The following question will however begin to get you to think about possible applications.

SAQ 2.12

You are working in a research laboratory attached to a large infectious diseases unit. The director asks you to develop a new therapeutic approach to a viral infection which is of epidemic proportions. The virus infects cells by binding to a T cell receptor we shall call X. You know very little about the structure of the virus. You have been asked to generate an immunotherapeutic agent that will prevent infection of the cells and destroy the virus. You are already aware that all patients infected with the virus have anti-viral antibodies in their serum but it is not protective. You have a supply of the cellular receptor. The soluble receptor does successfully prevent infection of the cells but is quickly degraded so is ineffective as an immunotherapeutic agent. Additionally, it does not kill the virus. Assuming that molecular genetics techniques are available to perform any modifications of your choosing which of the following is most likely to produce the desired result?

1) A murine monoclonal antibody to receptor X.

2) A human antibody to X.

3) A construct of receptor X attached to $F(ab')_2$.

4) A modified receptor X with a longer half life.

5) A molecule consisting of IgG antibody where each of the Fabs have been replaced by receptor X.

6) A human antibody to virus.

Summary and objectives

This has been a long chapter in which we have examined the structure and function of antibodies. We have examined the various classes of antibodies and introduced many of the terms used in describing antibodies. We have also shown how different parts of antibody molecules play different roles ranging from recognising antigen to bringing into action a number of common defence mechanisms.

Now you have completed this chapter you should be able to:

• list the products of enzyme digestion and reduction of antibodies;

• list the detailed structural and biological features of antibodies and identify structural and biological differences between the antibody isotypes;

• describe the special properties of IgA and its role in the secretory immune system;

• list the major features of the antigen combining sites and define the terms hypervariability, complementarity determining regions, framework regions, non-covalent interactions, specificity, affinity and avidity;

• define the conditions for complement activation and describe the steps that result in complement mediated lysis, chemotaxis, opsonisation, immunoadherence and phagocytosis;

• define the term protective antibody and list the required characteristics;

• explain the term heterogeneity and formulate protocols for the production of antibodies to isotypes, allotypes and idiotypes;

• propose modifications to the basic structural skeleton of antibodies to adapt them to immunotherapeutic functions.

Molecular genetics of antibodies

Molecular genetics of antibodies

3.1 Introduction

In the 1950's it was known that the amino terminal ends of antibody light and heavy chains comprising the V regions consisted of highly diverse sequences whereas the remainder of the molecule comprising the constant region was relatively invariant. It had also been concluded that the constant region was the product of a single gene due to the presence of allotypes on certain isotypic antibodies that were inherited in a Mendelian fashion. If there was a single gene for each antibody heavy or light chain then it would be necessary to propose that the part of the gene coding for the amino terminal variable region must have been subject to mutational changes during evolution to produce the necessary diversity seen in antibodies whereas the remainder of the gene coding for the constant region was somehow protected from such alterations.

The problem was solved in 1965 when Dreyer and Bennett suggested that there were two genes coding for each antibody chain, one for the variable portion and one for the constant region. This then lead to the idea of DNA rearrangement during development ie that one of the variable genes became joined to the constant gene at the DNA or RNA level finally giving rise to the complete antibody chain.

In this chapter we will describe how antibody gene rearrangement was discovered and examine our current knowledge on how antibody genes are arranged and expressed. In the final part of the chapter, we will examine some molecular genetic approaches for extending the use of antibodies for therapeutic purposes.

3.2 Discovery of antibody gene rearrangement

Southern
blotting

The actual proof that antibody genes are rearranged was provided in 1976 when Tonegawa reported on differences in the arrangement of the variable (V) and constant (C) genes on the chromosomes of embryonic and mature B cells using the Southern blotting technique (Figure 3.1). Embryonic cells are not committed to the production of any particular antibody whereas mature B cells can only produce one antibody of a single specificity. The mature B cells in this experiment were antibody-producing tumour cells from a plasmacytoma producing κ-light chains.

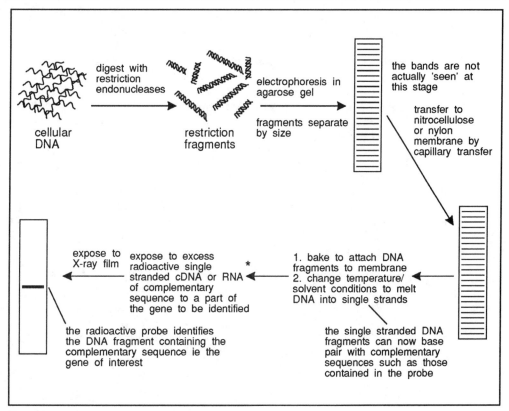

Figure 3.1 The basis of Southern blotting. * Tonegawa used mRNA. These days we use complementary (copy) DNA (cDNA) probes derived from RNA by reverse transcriptase.

restriction endonuclease

The DNA was extracted from both types of cells and subjected to digestion using the restriction endonuclease *Bam*H1. This bacterial enzyme recognises the base sequence GGATCC that appears at irregular intervals throughout the genome of an individual. The DNA was cleaved at these restriction sites and the resulting DNA restriction fragments ranging in size from about 0.5 kilobases to 10 kilobases were size separated in an agarose gel. The separated fragments were then transferred to a nitrocellulose membrane, baked to promote attachment of the fragments and then melted using changes in temperature and solvent into single stranded DNA. The membrane was then flooded with an excess of either one of two κ-mRNA probes labelled with radioactive iodine. One probe consisted of the complementary sequence for both variable and constant domains of the light chain and the second probe consisted of the 3' end of κ-chain mRNA only (the constant domain). The location of the hybridised probes was then detected by autoradiography.

hybridisation

Tonegawa found the pattern of hybridisation to be completely different in the genomes of the two cell types. The pattern of embryonic DNA showed two major bands when hybridised with the whole mRNA probe. One band, of molecular weight 6 million Daltons, also hybridised with the probe representing the 3' half mRNA. This suggested that this fragment contained the gene for the constant domain. The other band representing a smaller fragment of molecular weight of about 4 million did not hybridise with the 3' half mRNA and thus contained the V gene segment. This result suggests that the two gene segments were separated by a long stretch of DNA in embryonic cells that included at least one *Bam*H1 restriction site leading to their separation into two components on treatment with the enzyme.

In direct contrast, Tonegawa found that the pattern of mature B cell DNA showed only a single radioactive band that hybridised to both probes and was of lower molecular weight than either component derived from embryonic DNA (molecular weight of 2.4 million Daltons). This indicated that the V and C gene segments were on the same fragment of DNA and were no longer separated by a long stretch of DNA containing a *Bam*H1 restriction site.

DNA rearrangement

somatic recombination

Tonegawa interpreted these results to mean that V_κ and C_κ genes are some distance from each other in embryonic cells but are brought much closer together during differentiation of B cells committed to antibody synthesis. This process is called DNA rearrangement or somatic recombination. You will notice that a large section of DNA must have been excised in the process by comparing the molecular weights of the fragments and the absence of *Bam*H1 restriction sites between V and C genes in the mature B cell (Figure 3.2).

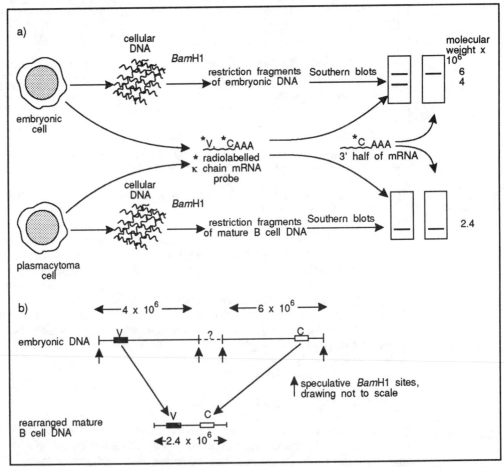

Figure 3.2 Tonegawa's experiment on DNA rearrangement. a) The experiment on DNA rearrangement. b) Diagrammatic interpretation of the results.

<table>
<tr><td>

SAQ 3.1

</td><td>

Southern blotting analysis of *Eco*R1-restricted sperm (embryonic) and plasmacytoma (myeloma 1 and 2) DNA hybridised with C_κ mRNA gave the results shown in Figure 3.3. Choose the correct explanations from those four given for each of the patterns observed in sperm, myeloma (1) and myeloma (2). Remember C_κ means the constant region of the κ chain.

1) Sperm cell DNA is of lower molecular weight than mature B cell DNA.

2) The kappa genes have rearranged on both chromosomes.

3) The kappa genes have rearranged on one chromosome.

4) *Eco*R1 did not have a restriction site between the V and C genes.

5) The band represents unrearranged DNA.

</td></tr>
</table>

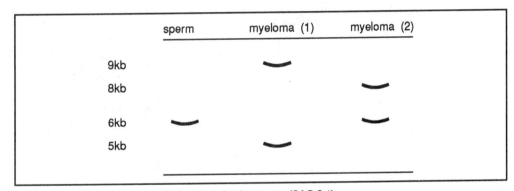

Figure 3.3 Southern blotting of *Eco*R1 restriction fragments (SAQ 3.1).

3.3 Antibody gene organisation

3.3.1 General features

unlinked gene families

There are unlinked gene families encoding the kappa (κ) and lambda (λ) light chains and the H chains. The H chain genes are found on chromosome 14 in Man and chromosome 12 in mouse; the corresponding chromosomes for the κ chain genes are 2 and 6 and for the λ chain genes, 22 and 16. The variable and constant genes for each chain are found near together on the same chromosome as suggested by the experiment described above and the basic arrangement of the genes in each set is very similar.

We should mention a practice among molecular geneticists which can lead to confusion. They always refer to the sequence of the non-transcribed strand of DNA which is of 5′ → 3′ direction rather than the transcribed strand which is 3′ → 5′. The reason for this is that this is the sequence found in the mRNA with uracil replacing thymidine. We shall adhere to this practice so that it will be less confusing to you when you read the literature. We must also remind you of two other terms. Exons are sequences of nucleotides in DNA that are transcribed and retained in RNA and are subsequently used to encode the amino acid sequences found in proteins. Introns are sequences of nucleotides in DNA that are transcribed but are subsequently removed from the RNA by RNA splicing. Thus these sequences are not translated into amino acid sequences (fuller details of RNA processing are given in the BIOTOL texts 'Infrastructure and Activities of Cells' and 'Genome Management in Eukaryotes').

immunoglobulin (Ig) locus
multiple V
D (diversity) segments
J (joining) segments
introns

The immunoglobulin (Ig) locus is composed of 3 - 4 regions. At the 5' end of the locus there are multiple leader (L) exons and V exons or segments. These V exons are not all identical but contain the full repertoire of V regions that the organism can express. Downstream from these are a collection of D (diversity) segments found in heavy chain genes only, further downstream are the J (joining) segments and at the 3' end of the locus are the C (constant) segments. Each exon or segment is separated by DNA of various lengths called introns which are noncoding sequences. The lambda light chain genes have a slightly different arrangement (see below).

DNA rearrangement

D segment

The mechanism of DNA rearrangement brings together one of the many V segments, a D segment and a J segment to provide a continuous message of VDJ in heavy chain genes and a V with a J to form a VJ message in light chain genes. During rearrangement there is a deletion of DNA segment. For example in heavy chains, D-J joining results in deletion of any D segments downstream from the D segment involved in the D-J join. We may represent this by:

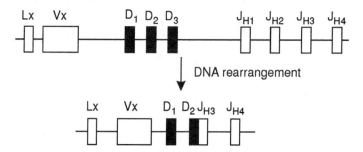

These rearranged gene segments then encode the whole variable domain of about 110 amino acids in heavy and light chains respectively.

So you can see that there is the potential in each individual, using DNA rearrangement, to create a much greater number of antibodies of different specificities by using different combinations of V, D and J segments than the total number of segments they possess. This mechanism is the basic method by which antibody diversity is generated. We will discuss this in more detail later.

leader exon

A short distance (90-200bp) upstream of each V exon is a so-called leader exon that encodes a translation initiation signal and most of a short leader peptide. This is found at the amino end of the translated Ig product and is involved in transfer of the Ig chain into the lumen of the endoplasmic reticulum where it is subsequently cleaved off and does not appear in the secreted product.

Each V segment encodes the last 4 amino acids of the leader peptide and the first 95 or so N-terminal amino acids of heavy and light chains that includes CDR1, CDR2 and framework regions (FR) 1-3. The D segments, when present, and the J segments contribute to the encoding of CDR3 and J also encodes FR4.

⫪ Based on the text so far, draw a diagram of the general chromosomal arrangement of the heavy chain genes with the following labels: intron, exon, leader exons, variable, diversity, joining and constant segments; orientation (5′, 3′ ends). Check with Figure 3.4. Read this figure carefully because if provides a lot of information we will discuss in the following sections.

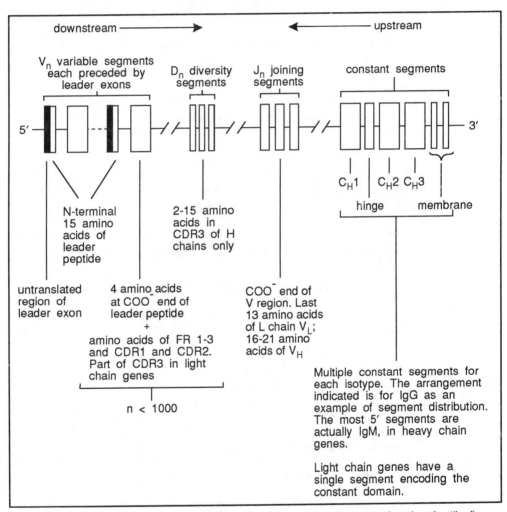

Figure 3.4 Immunoglobulin exons (segments) encoding the variable and constant domains of antibodies showing the general arrangement on the chromosomes. Notice that D segments are only in heavy chain genes and the arrangement of segments in lambda (λ) light chain genes differs from this arrangement. The values of n are only estimates based on Ig genes of man and mouse and includes nonfunctional genes (pseudogenes). Note that exons are represented by boxes, introns by a single line.

3.3.2 Lambda light chain genes

As we have said, there are no D segments in light chain genes and the whole variable domain is encoded by one V and one J segment. The V_λ segment encodes the first 97 amino acid residues of the lambda light chain and J_λ encodes the last 13 amino acid residues of the variable domain, both segments contributing to the CDR3.

Lambda light chain genes are unique in possessing very few V and J segments resulting in only limited diversity in lambda light chains and in the arrangement of the V, J and C segments in families in the genome.

Inbred mice only possess two V_λ segments. As shown in Figure 3.5 starting from the 5′ end of the λ locus, we find $V_\lambda 2$ with its leader exon, followed by 2 J-C pairs $J_\lambda 2$ - $C_\lambda 2$ and $J_\lambda 4$ - $C_\lambda 4$. Further downstream we come across the second V segment with its leader exon $L_\lambda 1$-$V_\lambda 1$ and further downstream 2 more J-C pairs $J_\lambda 3$, $C_\lambda 3$ and then, at the 3′ end of the locus $J_\lambda 1$, $C_\lambda 1$. Introns of about 1-3 kbp separate the J and C segments and an unknown distance separates the V and J segments.

pseudogenes   It is thought that $J_\lambda 4$ and $C_\lambda 4$ are pseudogenes and are not expressed. In DNA rearrangement, the V segments can only pair with the J-C clusters immediately downstream. Hence $V_\lambda 2$ can only pair with $J_\lambda 2$ and the resulting $V_\lambda 2 \, J_\lambda 2$ will use $C_\lambda 2$ for the constant domain. $V_\lambda 2$ cannot join with either $J_\lambda 3$ or $J_\lambda 1$. Similarly $V_\lambda 1$ will rearrange with either $J_\lambda 3$ or $J_\lambda 1$ and then choose $C_\lambda 3$ or $C_\lambda 1$ respectively.

In Man, an unknown number of V_λ segments are grouped upstream from 6 sets of JC pairs as shown in Figure 3.5. Presumably V segments are not restricted as in mouse and can rearranged with any of the 6 J segments. After V-J joining the corresponding C of the JC pair is used to encode the constant domain.

The 3 murine lambda constant segments exhibit considerable homology. The encoded products of $C_\lambda 2$ and $C_\lambda 3$ in mice differ in only 5 amino acid residues but both differ by about 40 amino acid residues from $C_\lambda 1$.

3.3.3 Kappa light chain genes

There are about 250 V_κ segments (possibly more) in mouse and 100 or less V_κ segments in Man. It may be that quite a number of the human V_κ segments are pseudogenes. They are arranged sequentially at the 5′ end of the kappa light chain locus, each preceded by a leader exon. The leader exons and V_κ segments are each separated by a 200bp intron. Some distance downstream there are 5 J segments in both mouse and Man, $J_\kappa 3$ in mouse being a pseudogene. Of 5 J segments in rabbits only one, J2, is known to be expressed. These J segments are separated by introns of about 300bp. About 2.5kbp downstream from the most 3′ J segment is a single C segment (Figure 3.5). Each V_κ encodes the first 95 amino acid residues of the kappa light chain. Both V and J encode parts of CDR3; in mice each J segment encodes two residues of CDR3.

SAQ 3.2	Which one of the following is correct? 1) Introns comprise a major portion of the eukaryotic genome and code for the protein products of the cell. 2) An embryonic B cell has constant and variable genes for each chain on separate chromosomes. 3) DNA rearrangements lead to VDJ joining in light and heavy chain genes. 4) Mature B cell DNA invariably contains less variable and diversity exons than germline DNA. 5) There are only 5V segments in lambda chain genes of mammals arranged in two families.

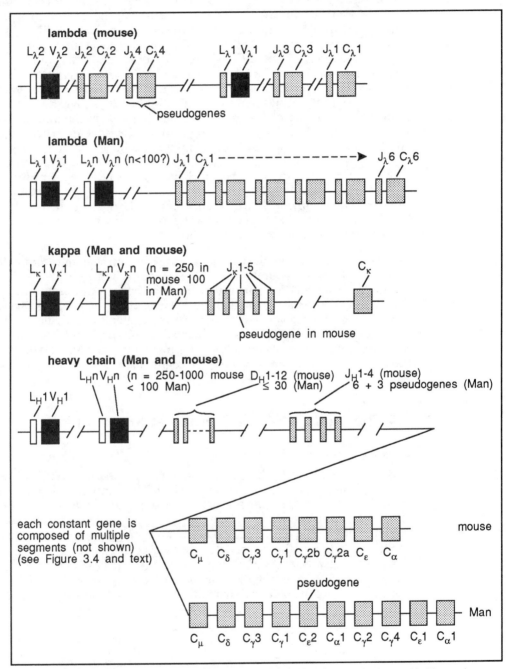

Figure 3.5 Organisation of germline Ig genes at the lambda, kappa and heavy chain loci in Man and mouse. Figure not to scale.

3.3.4 Heavy chain genes

Igh locus At the 5′ end of the heavy chain locus, called the Igh locus, is a cluster of 100-1000 V_H segments in mouse of which some 30% may be nonfunctional. In Man, the number of V_H segments is unlikely to exceed 100. As well as coding for the last 4 amino acid residues of the leader exons, the V segments encode the first 98 or so amino acids of the variable domains of heavy chains up to the end of the third framework region (FR3). An unknown distance downstream from the V_H cluster are 12 D_H segments in the mouse, possibly up to 30 in Man. D_H segments encode a major part of CDR3 in heavy chains, sometimes the whole CDR3 and vary considerably in length. Some D_H segments have been found to code for just 2 amino acid residues whereas others have been shown to encode 11 or more residues. You should note that there is a corresponding variability in the extent of the CDR3 sites in different antibodies. Downstream from the diversity segments is a cluster of J segments, numbering four in the mouse and nine in Man; however, three of the latter are thought to be pseudogenes. The J segments code for 16-21 amino acids compared to the 13 in light chain genes.

∏ As far as we know, in heavy chain genes any D segment can join with any J segment and any V segment can join with any D segment. Calculate how many potential variable domains can be encoded from these rearrangements in mouse heavy chains. Write your answer down. You will be able to check your answer at the beginning of the next section.

constant segments
6-8 kb downstream from the J cluster are the constant segments coding for the constant domains of the various antibody isotypes. Since the heavy chains of each isotype consist of multiple constant domains rather than the one domain found in light chains, the constant genes are more complex. For instance, there are 4 segments encoding the IgG heavy chain after the variable domain; three segments encode the 3 constant domains and another segment encodes the hinge region. For membrane receptor antibodies there are extra segments encoding transmembrane and cytoplasmic portions of the heavy chains. The arrangement of the segments for all the isotypes are indicated in Figure 3.5.

3.4 The basic DNA rearrangement mechanism

3.4.1 Potential for diversity

In the light chain genes there is a single joining event involving a V and J segment. Although there are restrictions imposed on V segments in lambda chain genes as mentioned already, there are no such restrictions for kappa and heavy chain genes. In other words, any V_κ can join with any J_κ and any V_H can join with any D_H which, in turn, can join with any J_H. Based on the germline in the mouse, then, there are potentially about 1000 kappa chains bearing different variable sequences ie 250 V x 4 J = 1000; and 48000 heavy chains ie 1000 V x 12 D x 4 J = 48000. We have already said, of course, that some of the V_H segments may be nonfunctional. As we shall see later this is a major underestimate of the potential diversity of antibody molecules.

SAQ 3.3	Let us assume that mice possess 1000 V_H segments and have 12 D_H and 4 J_H segments. The V segments each encode 98 amino residues. If we assume that D segments each encode, on average, 5 amino acid residues and J segments 18 residues, what percentage of the total mammalian genome is used to encode these segments? Assume that the total mammalian genome is 3.5×10^9 base pairs. Remember that each amino acid is coded for by three nucleotides and that mammals are diploid. Select your answer from those provided below.

1) 17%.

2) 0.005%.

3) 5%.

4) 0.008%.

5) 2%.

You should realise that all cells other than mature B cells possess Ig genes in the germline configuration as seen in embryonic B cells. It is B cells alone that are able to rearrange these genes by a process known as DNA rearrangement or somatic recombination that can only then result in the secretion of functional antibodies.

Let us examine the rearrangement mechanism for the heavy chain genes. Refer frequently to Figure 3.6 that illustrates this mechanism.

3.4.2 The D-J joining step

The first rearrangement leads to the joining of any D with one of the J segments. Let us say that there are 12 D segments numbered sequentially 1 - 12 ($5' \rightarrow 3'$) and 4 J segments numbered 1-4 ($5' \rightarrow 3'$) and that the developing B cell chooses to join D4 with J2. The introns and segments D5-12 and J1 will be excised or deleted and the remaining D and J segments will be D1-3,D4J2,J3,J4. This D4J2 joining step may be acceptable to the B cell and it may proceed to the second rearrangement where one of the V segments joins with the rearranged D4J2. However, the cell may proceed with a second D-J joining step. This time it may join D2 to J4 resulting in the deletion of the first rearrangement and the remaining D and J segments will now be D1,D2J4. In this particular example, no further rearrangement is possible since the cell has run out of J segments.

3.4.3 V to DJ joining

The second event is the joining of any of the V segments with the rearranged DJ resulting in a VDJ rearrangement. As in the DJ joining, the intervening DNA is deleted. This means that all the diversity segments have been deleted as well as the V_H segments downstream from the V_H chosen for the joining. This VDJ now encodes the sequence of the variable region of the antibody heavy chain product to be produced by this B cell and its progeny. The VDJ is still separated from the constant region genes by a fairly large intron which will contain all the remaining J segments downstream from the J involved in the joining step and the leader exon is still separated by a short intron from the rearranged V segment.

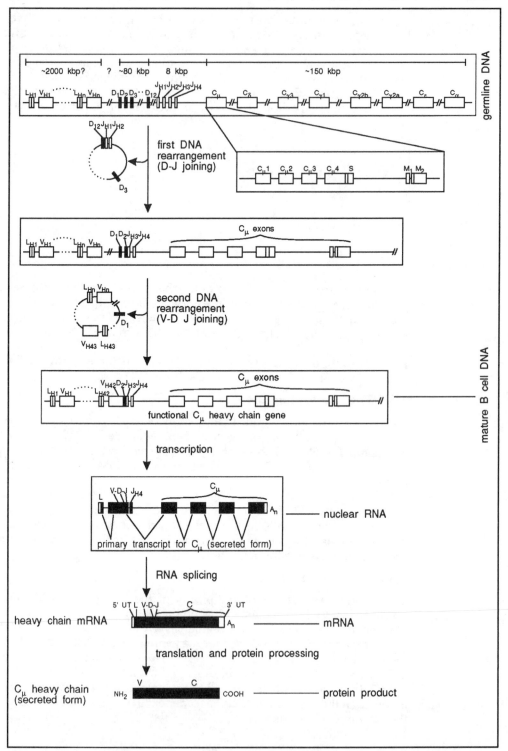

Figure 3.6 DNA rearrangement at the mouse heavy chain locus. Redrawn from Watson, Hopkins, Roberts, Steitz and Werner (1987), Molecular Biology of the Gene, Figure 23.26, Benjamin Cummings, California (see text for a description).

3.4.4 The primary RNA transcript and mRNA

primary RNA transcript

Subsequent transcription of the DNA into the primary RNA transcript in the nucleus is followed by polyadenylation of the RNA at the 3' end and the splicing out of the intervening intron separating the VDJ from the constant gene and the leader-V intron resulting in a functional mRNA for the IgM heavy chain. There may also be untranslated portions at each end of the resulting mRNA molecule consisting of uracil and thymidine (5' UT and 3' UT). It is thought that the addition of poly-A tails is necessary for processing of the RNA. The pre-B cell now produces the IgM heavy chain which is retained in the cytoplasm after cleavage of the leader peptide.

IgM heavy chain

The production of cytoplasmic μ chain is thought to signal a productive rearrangement and somehow inhibits further rearrangements of heavy chain genes in the B cell. However, recent evidence suggests this may not be so and regulation of rearrangements remains to be elucidated.

nonproductive rearrangement

Only about one third of somatic recombination events are successful. If the VDJ rearrangement described above results in a nonproductive rearrangement, then the Ig locus on the other chromosome (allele) goes through the same process. If this too is unsuccessful, the result is a nonfunctional B cell. However, recent reports suggest that this is still not the end of the story for these B cells (see Section 3.8).

Π   To ensure that you are now familiar with the rearrangement mechanism, draw a diagram of the following rearrangement in a similar manner to Figure 3.6. Assume there are segments V_H numbered (5' → 3') 1-12, D_H segments 1-12, J_H segments 1-4 and a single constant segment. Aim for a rearranged V gene of $V_H6,D4J_H2$ and include in your diagram the deleted sequences at the DNA level and in the spliced out primary transcript. You should be able to check it against Figure 3.6.

3.4.5 Light chain gene rearrangements

Following the successful V_HDJ_H rearrangement on either chromosome, the pre-B cell will now attempt VJ joining at the kappa locus. If successful on either chromosome, the B cell will synthesise kappa chains and will express membrane IgM ($\mu_2 \kappa_2$) on the cell surface. If both kappa rearrangements are nonproductive then the cell will rearrange at the lambda locus. If successful, the cell will express membrane IgM consisting of heavy chains and lambda chains. If nonproductive, the cell presumably dies (Figure 3.7). Follow the sequence carefully.

Once the B cell leaves the bone marrow and makes contact with antigen, secretory IgM will be produced. Sometime later in the life of the B cell it may switch from making IgM to producing another isotype (Section 3.9).

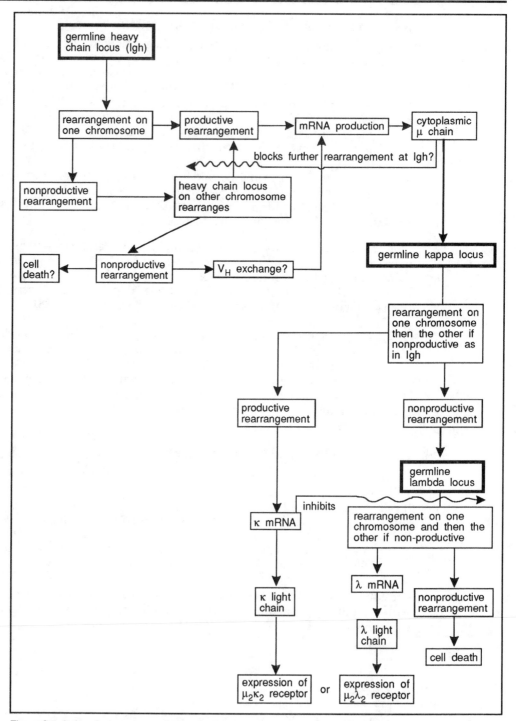

Figure 3.7 Ordered rearrangement of Ig genes.

3.4.6 Allelic and isotype exclusion

allelic exclusion   This series of events should give you some basis for the phenomenon of allelic
                      exclusion. For heavy chains, a productive rearrangement of VDJ will presumably lead

to the use of the constant segments of the immunoglobulin isotypes on the same chromosome ie either the maternal or paternal chromosome. If the initial VDJ rearrangement is productive, the synthesis of the IgM chain blocks rearrangement on the second chromosome. This is called allelic exclusion.

isotype exclusion

Since lambda chain genes are only rearranged if the kappa rearrangements are nonproductive, this explains why a single B cell can only produce one of the two types of light chain and why antibody molecules are never light chain hybrids. This phenomenon is called isotype exclusion.

SAQ 3.4

Which one of the following is a possible murine rearrangement?

1) $V_H4D2 \rightarrow V_H4D2J_H2$

2) $D_\lambda2J_\lambda4 \rightarrow V_\lambda2D_\lambda2J_\lambda4$

3) $V_\lambda2J_\lambda2 \rightarrow V_\lambda2J_\lambda2C_\lambda4$

4) $D2J_H3 \rightarrow V_H4D2J_H3$

5) $V_\kappa3J_\kappa2 \rightarrow V_\kappa3J_\kappa2C_\kappa2$

3.4.7 Recombinase enzymes promote DNA rearrangements

recombinases

Exon joining probably occurs by excision of the DNA between the exons by a 'looping out' process. A system of enzymes called recombinases are responsible for this process although it has to be admitted that we know little about their action. Two genes have recently been identified that stimulate somatic recombination in Ig genes. These have been called recombination activating genes RAG 1 and RAG 2 but the nature of their action is not known. The main features of the recombinase enzymes that you need to know are as follows:

recombination activating genes

• They only act in immature or developing lymphocytes thus ensuring that Ig gene rearrangements do not proceed in other cell types that also possess the Ig loci. Notice, however, that they are also present in T lymphocytes since T cell receptor genes also undergo an analogous process;

• Recombinases appear to recognise unique sequences associated with each of the exons comprising the variable genes. We shall now examine this mechanism in more detail.

3.4.8 Nonamers, heptamers and spacers

conserved heptameric and nonameric sequences

Immediately downstream (3′) from each V segment, upstream (5′) from each J segment and situated on (flanking) both sides of D segments are highly conserved heptameric and nonameric sequences separated by either a nonconserved 12 ($\pm$1) or nonconserved 23 ($\pm$2) nucleotide spacer. It would appear, then, that the actual nucleotide sequences of the heptamers and nonamers are important for recognition whereas the nonconserved nucleotide spacers are important for their length. It is, perhaps, of interest that 12 and 23 nucleotide spacers represent one and two turns of a double helix respectively.

nonconserved nucleotide spacers

On examination of these sequences and spacers we would also notice that their orientation is important in dictating which exons can join with what. There are joining rules that simply state that:

joining rules

- exons may join only if they are abutted by spacers of opposite types (12/23bp rule);

- they need complementary (inverted repeats) heptamer and nonamer sequences.

By inverted repeats we mean sequences of nucleotides that are found in the opposite orientation elsewhere in the DNA.

These rules are best illustrated by a couple of examples.

Use Figure 3.8 to help you follow the descriptions of these examples.

1. In kappa chain genes, each V segment is followed by a heptamer, then a 12 nucleotide spacer and finally a nonamer sequence. Each J_κ segment is preceded immediately upstream by the complementary or inverted repeat heptamer sequence, 5' to this is a 23 nucleotide spacer and finally the complementary nonamer sequence. Since these arrangements adhere to the joining rules given above, V_κ are able to rearrange with J_κ segments.

It is interesting to note that in lambda chain genes the spacer arrangement is opposite ie the V segment has a 23 nucleotide spacer and the J segment a 12 nucleotide spacer. So V_κ segments cannot join to J_λ segments. Since kappa and lambda chain genes are on separate chromosomes this does not appear likely but such arrangements could have some significance in the expression of the two light chain genes (see Figure 3.7).

2. The situation is more complex with heavy chain genes as there are two somatic recombination events and they must proceed in the correct sequence. At the heavy chain locus, each V segment is followed downstream by a heptamer, then a 23 nucleotide sequence and finally a nonamer. Each V segment must be regulated so that it only joins to a rearranged DJ segment at the 5' end ie to a D but not to a J_H segment. Thus we find that upstream from each J segment there is the same heptamer sequence (not the inverted repeat) as for the V segment preceded 5' by a 23 nucleotide spacer (again, same as V) and finally by the same nonamer sequence. Based on the joining rules then, V cannot join with J. To accommodate joining with both J and V segments, D segments must have the same arrangement both upstream and downstream ie first, inverted repeat sequences of the heptamer, then 12 nucleotide spacers and finally inverted repeats of the nonameric sequence and this is indeed what we find in the heavy chain locus.

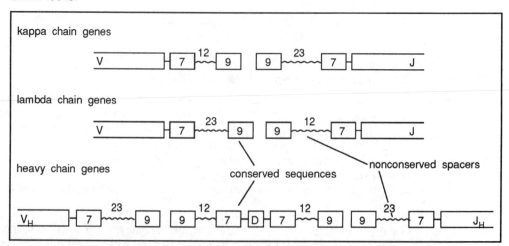

Figure 3.8 Organisation of the recognition heptameric and nonameric sequences and 12 and 23bp spacers at the kappa, lambda and heavy chain loci.

3.4.9 Model for the joining mechanism

We do not really know how the conserved sequences and nonconserved spacers are used to align the segments to be joined but there are a few suggested models, the most plausible is shown in Figure 3.9. In this model, the complementary heptameric and nonameric sequences base pair to form a stem structure also containing the apposing 12 and 23 nucleotide spacers. It is assumed that the recombinase enzyme complex recognises this whole structure. Notice that the intervening segments and introns are contained in a loop and subsequent excision would remove both the stem and loop resulting in the joining of the two segments. Presumably the same event occurs simultaneously on the complementary strand of DNA. Notice that, on being joined by the recombinase enzymes, the intervening DNA is excised (deleted).

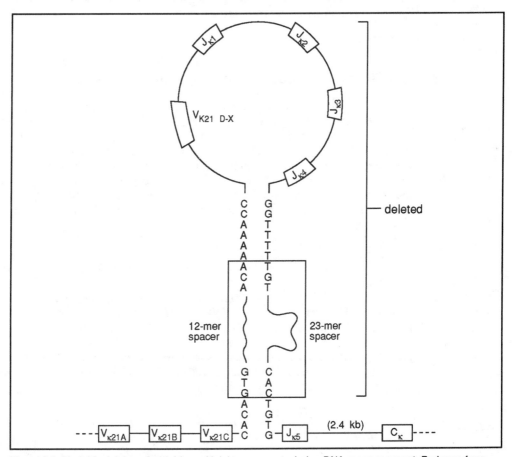

Figure 3.9 Model for joining of variable and joining segments during DNA rearrangement. Redrawn from 'The Experimental Foundations of Modern Immunology' by William R Clark John Wiley & Sons Ltd 4th Edition 1991 page 106 Figure 5.8.

Which one of the following is correct?

1) Spacers, nonamers and heptamers are conserved recognition sequences.

2) V_H gene formation requires a single DNA rearrangement.

3) There is no joining of variable and joining segments because they are abutted by the same 23 base pair sequences.

4) V_H encodes the first 3 framework regions and two CDR regions.

5) When joining between segments occurs, the heptameric sequences are retained and not deleted.

3.5 Diversity based on the germline genes

The diversity of the antibody pool of an individual is primarily based on the possession of multiple germline V genes for both heavy and light antibody chains. Increased diversity is generated by the utilisation of the D and J segments but only at the third hypervariable region, CDR3. Since man possesses no more than 100 V_H segments, there is limited V_H diversity and repeated choices of a particular V_H segment is translated into the same CDR1 and CDR2 expressed in the antibody product. It is apparent, therefore, that a major contribution to diversity and, indeed antigen binding, is by CDR3. As we have seen, CDR3 is, in large part, encoded by the J and/or D segments.

limited V_H diversity

You may remember that in Chapter 2 we discussed the use of kappa and lambda light chains by different species. Most antibodies in mice carry the kappa chain. This is of obvious advantage to the mouse since there are many kappa V segments but only two lambda V segments. Obviously, antibodies containing lambda chains are going to be of more restricted specificity. Man, on the other hand, produces more antibodies containing lambda chains (about 40% of all antibodies) and this correlates well with a larger number of lambda V segments in the human genome.

Additional diversity is generated by the random association of heavy and light chains. This assumes that there is a random selection of light chain genes following productive rearrangement of the heavy chain genes.

Replace the bold words or numbers with the correct ones in the following statements:

1) Each V segment encodes the last **7** amino acids of the leader exon and the first 95 or so amino acids including the 3 CDR's and **4** framework regions.

2) Joining mechanisms involve the complementary base pairing of **hexameric** and nonameric sequences which are recognised by a **ligase** enzyme.

3) V segments in lambda chain genes have a **20** nucleotide spacer and the J segment a **23** nucleotide spacer. The arrangement on the chromosome downstream from each V segment is a **nonamer** followed by spacer then the **heptamer**.

3.6 Non-germline generation of diversity

non-germline
diversity

We have already calculated the potential diversity based on germline rearrangements. We will now examine ways in which alterations in germline sequences (non-germline diversity) can be introduced resulting in an even wider spectrum of antibody specificities.

3.6.1 Junctional diversity - imprecise joining

During V to J, V to DJ and D to J joining, there is an imprecision in the manner in which a particular nucleotide is chosen as the joining point. Hence, nucleotide sequences from the two joining segments can join at any of several nucleotides leading to slightly different base sequences distinct from the germline sequence of the two segments junctional involved in the joining. This results in what is called junctional diversity. If the resulting diversity recombination is still in frame, the DNA can be transcribed and the resulting RNA transcript will encode an Ig chain with an amino acid substitution in CDR3. Due to the degeneracy of the genetic code, however, junctional diversity does not always result in such changes. This is explained in Figure 3.10.

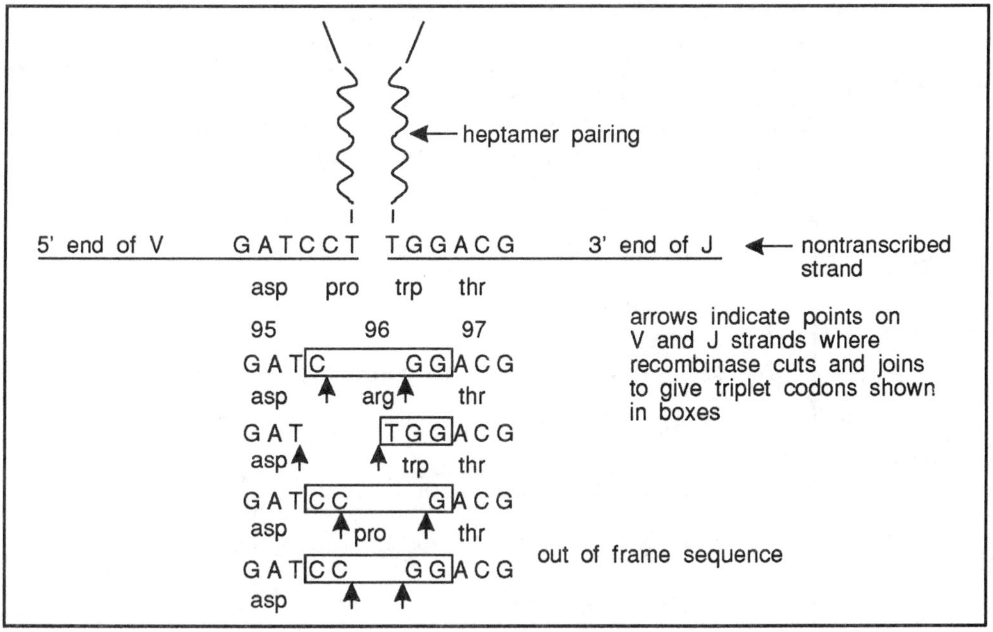

Figure 3.10 Junctional diversity resulting from imprecise joining of V and J segments at codons for amino acid residues 95 - 97 (within CDR3).

As Figure 3.10 indicates, this can also lead to out of frame rearrangements so that the DNA cannot be transcribed; sometimes even a stop codon may be inserted. Junctional diversity obviously contributes to nonproductive rearrangements. However, B cells can deleting compensate for out of frame joinings by deleting nucleotides 5′ of the joint bringing the nucleotides sequence back into frame. Nucleotide sequences (N segments) may also be inserted so the final sequence in the CDR3 may bear only a slight resemblance to the prototype gene segment.

3.6.2 Junctional diversity - N segments

N segments

As we mentioned above, additional diversity is also generated by the insertion of non-germline nucleotide sequences at the V-D and D-J junctions during rearrangement of heavy chain genes. These are called N segments and may extend to as many as 30 nucleotides resulting in a much extended CDR3. It is thought that they are added by the enzyme terminal deoxyribonucleotide transferase.

These two mechanisms obviously result in considerably more CDR3 sequences in antibody molecules than the total number of V, D and J exons in the germline and once again emphasises the importance of this hypervariable region to antibody specificity.

3.6.3 Gene conversion

gene conversion

We should just refer to the use of V segments in chickens because there is only one functional light chain V and J segment. However, there are many V pseudogenes and these appear to provide short stretches of nucleotides that are exchanged with similar stretches within the functional V segment. The mechanism of gene conversion has not been worked out.

3.6.4 Novel rearrangements at the D locus

novel rearrangements

Recent research reports have established that the 12/23bp rule is not absolute. For instance, rare cases of rearranged heavy chain V-J, V_H-D -D J and 3'D 5'-J_H (inverted D-J joining) as well as other novel rearrangements have been found and some of these also involve simultaneous N segment additions sometimes exceeding fifteen nucleotides in length. It is beyond the scope of this unit to postulate how these come about. It is sufficient that you know they exist and that they provide further opportunities for the generation of diversity.

SAQ 3.7

Which one of the following is correct?

Junctional diversity

1) only occurs in V-J joining in light chain genes.

2) can lead to stop codon generation.

3) occurs in all combinatorial events involving CDR3.

4) involves the 3' end of the V segment.

5) always involves N segments.

3.6.5 Somatic mutation

somatic mutation

All the mechanisms discussed so far to generate diversity operate in the absence of antigen in developing B cells. There is yet another mechanism to generate diversity that operates in mature B cells under the influence of antigen that results in further structural alteration of the nucleotide sequences of primarily V genes. This is called somatic mutation or hypermutation which is now thought to contribute greatly to the affinity maturation of antibodies during the immune response.

It was concluded that somatic mutation occurs when investigators discovered by analysing the nucleotide sequences of Ig genes in antibody producing cells that there appeared to be more V_H regions expressed in IgG producing cells in late primary and secondary responses than early in the primary response.

It has been reported from experiments in which nineteen V_H sequences encoding the first 100 or so N-terminal amino acids in phosphoryl choline antibodies were compared with germline sequences of four selected V regions known to encode specificity for the antigen. They found that all nineteen sequences were related to only one of the four prototype sequences called T15. Ten of the sequences were identical to T15 and were of the IgM isotype. The remaining nine sequences differed from the germline sequence by one to eight amino acid residues and all the variants were of the IgG of IgA isotypes. The mutations were distributed in both framework and CDR regions.

Π Before you read on, write down conditions or rules which you think may govern somatic mutation based on the above findings. Check with the text below.

Based on this and other evidence in the literature, we can make a number of observations on the phenomenon of somatic mutation.

- It is normally restricted to productively rearranged immunoglobulin V, D and J gene elements or within a 1000bp of it. Mutations are more common in the hypervariable regions. However, mutations have also been reported recently in nonproductive VDJ elements. Another report shows more mutations in the constant genes of monoclonal antibody cell lines and, more surprisingly, that such mutations changed the avidity of antigen binding! However, mutations are rarely, if at all, found in unrearranged genes;

restricted B memory cell population
- The mutations arise as a result of antigen stimulation. They are derived from germline V sequences and seem to be introduced one at a time and at different positions in a restricted B memory cell population during primary immunisation. In some reported experiments they are concentrated over a few days starting at about day 5 of the primary response. It has been recently suggested that memory B cells recruited into a secondary response do not undergo further mutation but this is controversial. Memory B cells that have undergone mutations resulting in greatly increased affinity for antigen are strongly selected for by antigen;

somatic mutations can generate new specification
- Mutations have been observed to greatly increase the affinity of an antibody. Antibodies against anti-oxazolone from B cells containing 14 mutations in the immunoglobulin producing gene have a 100 fold higher affinity for the antigen. Likewise the introduction of 7 mutations in the immunoglobulin gene of cells producing anti-nitrophenyl antibodies have been reported to have an affinity ten times that of the germline derived antibody. However, it should be obvious to you that mutations can also result in lower affinity and even loss of antigen binding altogether;

- IgM has a low mutational frequency but this is increased in class switched cells (cells which switch to making a different class of antibodies). Since class switching (see below) is often cytokine-dependent, T cell-derived growth factors (cytokines) could be involved in promoting mutations in V genes. This correlates with the finding of few mutations in T cell independent responses.

In conclusion, somatic mutations may be extremely important in affinity maturation of antibodies and may also be a means of generating new specificities.

SAQ 3.8

Here is the genetic code which you will need to use to be able to answer this SAQ.

5'-OH terminal base	Middle base				3'-OH terminal base
	T	C	A	G	
T	Phe	Ser	Tyr	Cys	T
	Phe	Ser	Tyr	Cys	C
	Leu	Ser	TC	TC	A
	Leu	Ser	TC	Trp	G
C	Leu	Pro	His	Arg	T
	Leu	Pro	His	Arg	C
	Leu	Pro	Gln	Arg	A
	Leu	Pro	Gln	Arg	G
A	Ile	Thr	Asn	Ser	T
	Ile	Thr	Asn	Ser	C
	Ile	Thr	Lys	Arg	A
	Met	Thr	Lys	Arg	G
G	Val	Ala	Asp	Gly	T
	Val	Ala	Asp	Gly	C
	Val	Ala	Glu	Gly	A
	Val	Ala	Glu	Gly	G

TC = termination codon

Examine the two sequences below. They are the 3' end of V kappa segment and the 5' end of a J segment. Find the heptameric sequences based on Figure 3.9 and then determine which two of the numbered amino acid sequences derive from somatic mutation rather than junctional diversity. Choose the one correct combination from (a to e).

V3'-TTG TAC CCT GTC AGA ACC ACA GTG TTT ACT - - -

- - - ACT CAC TGT GCA TAC TGC TAC TTC GAT-J5'

In the absence of junctional diversity the sequence is (V→J):

leu-tyr-pro-val-arg-tyr-cys-tyr-phe-asp

Sequence 1: leu-tyr-pro-val-ser-tyr-cys-tyr-phe-asp
Sequence 2: leu-tyr-pro-val-ser-cys-tyr-phe-asp
Sequence 3: leu-tyr-pro-val-arg-thr-tyr-cys-tyr-phe-asp
Sequence 4: leu-phe-pro-val-arg-tyr-cys-tyr-phe-asp
Sequence 5: leu-tyr-pro-val-arg-ala-tyr-cys-tyr-phe-asp

a) Sequences 1 and 3
b) Sequences 2 and 5
c) Sequences 3 and 5
d) Sequences 2 and 4
e) Sequences 1 and 4

3.7 V$_H$ gene usage and CD5 B cells

In this section we will be referring to cell markers (especially Lyl and CD5). These are two of many cell surface molecules expressed by various types of cells and can be used to identify the cells, hence the term markers. The markers are identified using antibodies. The CD classification refers to human cells and is described in Section 5.3.1, the Ly system refers to murine cells.

We will also refer to specific V$_H$ segments or families of V$_H$ segments which have been given code markers. We do not expect you to remember the codes or their relative positions in the V$_H$ gene cluster.

A minor percentage of B cells expressing the marker Ly1B in mice and CD5 in Man appear to develop early in development (ontogeny) and persist as mature B cells over long periods of time in the absence of any somatic mutation. They are present in most lymphoid tissues of adult animals and actually represent some 50% of total B cells in the peritoneal cavities of mice. They are the predominant B cell type in murine and human foetuses and may represent a separate B cell lineage to the Ly1$^-$ or CD5$^-$ B cells which represent the majority of B cells in Man and mouse.

separate B cell lineage

CD5 cells which may constitute up to 25% of the circulating and splenic B cells in human adults, are thought to be responsible for the production of polyreactive antibodies otherwise known as natural antibodies. Many of these have specificity for self antigens (autoantibodies) such as the Fc fragment of IgG (antibodies called rheumatoid factors), single stranded DNA, thyroglobulin and others. These polyreactive antibodies also possess a spectrum of specificities for common bacterial antigens such as polysaccharides and phosphoryl choline. The polyreactivity of these antibodies has been demonstrated. For example experiments have shown that the binding of one antibody to IgG Fc could be inhibited by single stranded DNA, tetanus toxoid, thyroglobulin and insulin. These antibodies may also include specificities for the major blood group antigens.

polyreactive antibodies

rheumatoid factors

We have kept referring to the random usage of any V segment in the rearrangement mechanism. Indeed, if there was random usage of the V gene pool we would expect that the pool of antibodies produced would be representative of all the V genes in the genome. Although this is correct when you examine adult CD5$^-$ B cells in, say, the spleen, it is certainly not so if you examine immature B cells or mature CD5 cells.

In immature cells then, whether CD5$^+$ or not, a preferential usage of the V$_H$ segments proximal to the D segments (ie the most 3' V$_H$ segments) has been suggested. For instance, 80% of a selection of murine foetal or neonatal cells were found to rearrange V$_H$7183 situated near the D cluster and V$_H$ Q52 at the 3' end of the V locus was also preferred. However, it is not yet clear what selective mechanism operates since some members of the family VGAM3.8 are nearest to the D segments and there is no difference in their usage in foetal and adult cells!

pre-immune
repertoire

These early cells are collectively called the pre-immune repertoire. This distorted usage has been found, rather surprisingly, in adult bone marrow B cells which presumably give rise to the adult B cells in the spleen and lymphoid tissue which express a more random pattern of V gene usage. However, CD5 cells maintain their restricted usage of the 3' proximal V_H genes throughout life. It is also interesting that mitogen stimulation of these adult bone marrow cells but not adult peripheral B cells results in autoantibody production. Additionally, B cells of mice with some autoimmune diseases show

autoimmune
diseases

preferential expression of 3' V_H genes and CD5 B cells are prominent in autoimmune disease. One last point is that CD5 B cells maintain germline expression of V_H genes and rarely undergo somatic hypermutation. There is a preponderance of IgM producers in this cell population although some IgG and IgA producers have been reported.

A non-random pattern of J segment usage also appears to operate in all B cells. One laboratory recently examined 111 VDJ sequences from 6 individuals and found that J_H4 was selected in 52% of the rearrangements whereas $J_H 1/2$ were only present in 1% of the sequences. There does not appear to be the same biased use of D segments.

3.8 A novel V_H to V_HDJ joining mechanism in pre-B cells and mature B cells

You may well be asking the question how do adult peripheral B cells manage to acquire a more random expression of V gene usage if they are derived from bone marrow cells that rearrange mainly 3' V_H genes.

As we mentioned earlier (Section 3.4.4) a pre-B cell can rearrange a VDJ on the second chromosome if the first rearrangement is nonproductive. If this second rearrangement is also nonproductive, then the cell is called a null cell because it cannot produce an IgM heavy chain. However, recent evidence now suggests that these cells are not dead-end products but can generate a new V_HDJ using a 5' V_H exon to replace the V_H sequence in the nonproductive V DJ. A similar mechanism appears to be operative in mature B cells. This could explain the switch to random expression of V segments in splenic B cells from the restricted 3' V usage of their bone marrow precursors. A suitable term for this

V_H exchange

is V_H exchange.

The ability to exchange V_H segments after a V_HDJ rearrangement appears to involve recognition of a heptamer embedded in the V coding sequence which is identical to the signal heptamer sequence found 5' of D segments. This heptamer is conserved in over 70% of all known V_H genes and is situated within a few nucleotides of the 3' end of the V_H sequence. It has been proposed that a looping out of the intervening DNA between

embedded
heptamer

the V_HDJ and the exchange 5' V segment occurs so that embedded heptamer aligns with the heptamer signal sequence 3' of the exchange V_H segment. This results in deletion of the rearranged V segment by the exchange resulting in deletion of the rearranged V segment by the exchange V. Figure 3.11 describes the proposed mechanism. This is one possible mechanism that could lead to random usage of V_H observed in the adult repertoire.

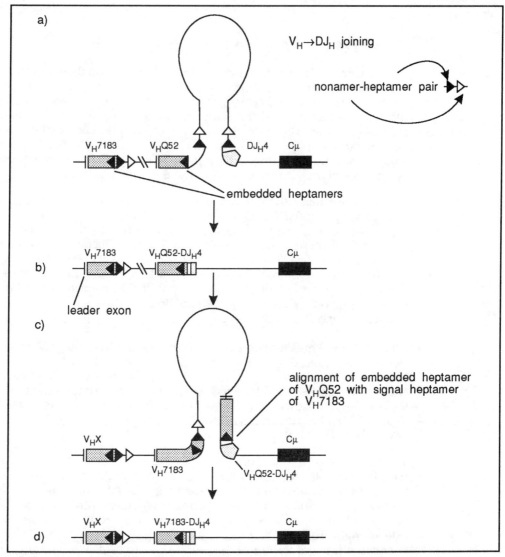

Figure 3.11 Possible mechanism for V_H exchange a) Initial $V_HQ52 \rightarrow DJ_H4$ joining to give a V_HQ52-DJ_H4 unit. c) Further re-alignment using embedded heptamer of V_HQ52 with signal heptamer with DNA excision leads to d) the production of V_H7183-DJ_H4. Adapted from Kleinfield R 'Recombination between an expressed immunoglobulin heavy chain gene and a germline variable gene segment in a Lyl[+] B cell lymphoma', Nature 322, 843-845.

SAQ 3.9

Match the items in the left column with those in the right column.

1) Somatic mutation

2) CACTGTG

3) CD5[+] B cells

4) Junctional diversity

5) Rheumatoid factor

a) V_H exchange

b) Fc region

c) CDR3

d) Germline genes

e) Immune response

3.9 Class (isotype) switching

isotype
switching

Early B cells are known to express both IgM and IgD on their surfaces. After antigenic stimulation, some of the clonal progeny switch their isotype and secrete, and also express on their cell surface, IgG, IgA or IgE. The antigen specificity of the cell remains the same indicating that the same rearranged V gene is being used. This isotype switching appear to be under the influence of T helper cells and their secreted cytokines (see Chapter 6) and, as we have already indicated, these also promote somatic mutation to higher affinity antibodies. Although the actual mechanism has not been defined there is strong evidence for two alternative molecular mechanisms.

switch regions

The first mechanism involves the deletion of heavy chain constant genes in a manner which may be similar to VDJ rearrangement. This is based on evidence from various myelomas producing a single isotype that have deleted all the constant genes 5' of the gene for that isotype. As an example, a myeloma producing IgG1 had deleted the genes for IgM, IgD and IgG3 (see Figure 3.5). The mechanism of switching appears to involve so-called switch regions which precede (5') each C_H gene cluster except C_δ. These regions occupy nucleotide stretches of up to 10 000bp consisting of tandem repeats of highly conserved sequences of up to 52bp long. It is not known whether there are unique sequences for each isotype.

When this mechanism operates, say for a class switch from IgM to IgG1 in man, the switch region preceding the IgG1 cluster is recognised by either a class specific protein or enzyme. The intervening DNA (containing the constant gene clusters for IgM, IgD, and IgG3) is deleted, and the IgG1 gene cluster now occupies the position vacated by the IgM cluster 3' down from the rearranged VDJ ready for transcription to IgG1 mRNA (Figure 3.12). This process is irreversible ie the cell can no longer produce either IgM or any of the deleted isotypes.

alternative
RNA splicing

The model (Figure 3.12) fails to explain the expression of IgM and IgD membrane receptors and the secretion of another isotype as found in some B cells. A second mechanism which explains this phenomenon is called alternative RNA splicing and leads to production of mRNA's for the simultaneous translation of more than one isotype (Figure 3.13). It is thought that a primary RNA transcript is produced that extends right from the VDJ through to the most 3' gene cluster being expressed. By alternate splicing of this transcript, mRNAs could be produced for production of more than one heavy chain isotype as shown in Figure 3.13. Using this mechanism, DNA for any of the gene clusters is not deleted. It is thought, however, that the cell finally irreversibly switches to production of a single isotype using the deletion method described above.

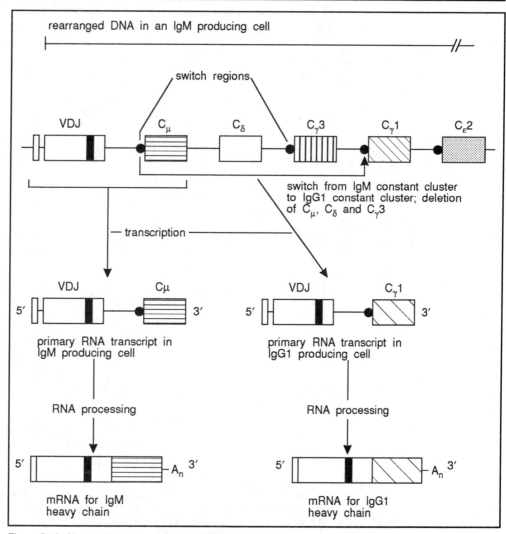

Figure 3.12 Class switching by deletion of DNA.

simultaneous
expression

Alternative RNA splicing also explains the simultaneous expression of cell surface IgM and IgD and the secreted products. A primary transcript is produced consisting of the rearranged VDJ as well as the C_μ and C_δ sequences. One of the constant gene sequences is spliced out of each of the transcription molecules resulting in mRNAs for both IgM and IgD heavy chains.

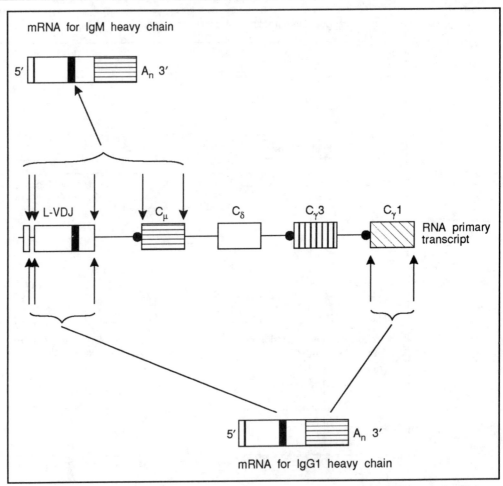

Figure 3.13 Class switch from IgM to IgG1 by alternative splicing of a long primary RNA transcript. The upper spliced RNA will lead to the production of IgM heavy chains. The one illustrated at the bottom of the figure will lead to the production of IgG1 heavy chains. ↑↓ splice points in RNA processing.

Π   It will be good practice for you now to go through the following exercise. You will need a large sheet of paper. Draw a diagram of the steps necessary to create the heavy chain gene $V_H D_H J_H C_\alpha 1$ in Man. We will assume (wrongly, of course) that there are 5 V_H, 4 D_H and 5 J_H segments. Include the original germline configuration, the organisation after VDJ joining and after class switching. Also include the primary RNA transcript, the mRNA after processing and the product sequence. Check your work against Figures 3.5, 3.6 and 3.12. This exercise provides excellent revision and will enable you to see how the various mechanisms we have discussed fit together.

3.10 The production of membrane and secreted forms of antibodies

The heavy chains of secreted and membrane IgM differ in amino acid sequence at the carboxyl terminal ends. Secreted IgM possesses a tail piece of 20 amino acids attached to the $C_\mu 4$ domain. In membrane IgM, the last domain is followed by an amino acid spacer, then a transmembrane portion of 26 hydrophobic amino acids and finally a cytoplasmic tail of 3 amino acids. The hydrophobic amino acids will bind the antibody to the membrane. We can represent this in the following way:

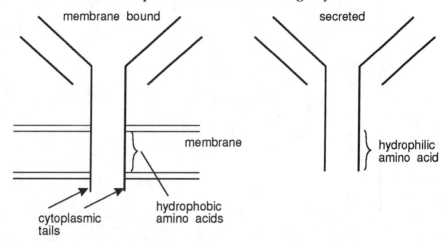

The DNA sequence within the C_μ cluster contains the information for both the secreted and membrane forms of IgM. Use Figure 3.14 to help you follow the description of switch from membrane bound to secreted IgM synthesis. The switch from one form to the other is again at the level of RNA processing. The IgM constant gene cluster consists of exons for the 5 constant domains, a short exon for the tailpiece attached 3' to $C_\mu 4$ and further downstream two exons for the transmembrane and cytoplasmic regions. The primary RNA transcript contains all of these and alternative processing determines which of the additional exons for the tailpiece or transmembrane and cytoplasmic portions are included in the mRNA. Hence, both membrane and secreted forms can be produced simultaneously (Figure 3.14). C_H genes for the other isotypes have similar exons for membrane and secreted forms of immunoglobulin.

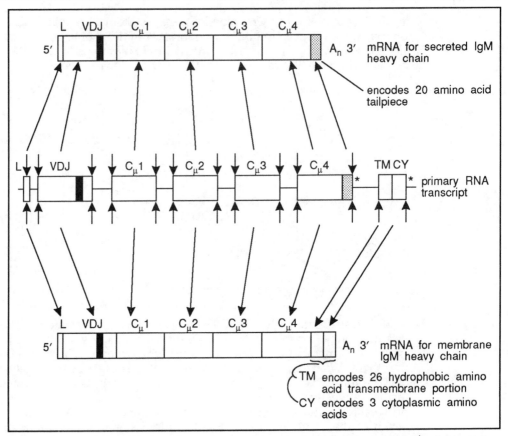

Figure 3.14 Alternative processing of RNA for membrane and secreted IgM heavy chain. ↑ splicing points in RNA processing. * polyadenylation sites.

SAQ 3.10	Which of the following is/are correct?

1) Class switching results in the same idiotype and allotype being expressed but with a different isotype.

2) Alternative RNA splicing results in the loss of immunoglobulin constant genes from the germline.

3) Human B cells can switch from production of IgM to IgG1 and subsequently to IgG4.

4) The heavy chain of membrane IgM is longer than that of secreted IgM.

5) All constant genes possess switch regions.

3.11 Regulation of immunoglobulin gene transcription

promoter
elements

TATA box

conserved
octamer

enhancer
sequences

Unrearranged V segments are not transcribed. Only after the V segment rearranges with DJ segments can transcription proceed. How is this regulated? Well, there are promoter elements lying a short distance upstream of each leader segment. These include a so-called TATA box (an AT rich sequence common in eukaryotic promoters) which is situated about 25bp upstream from the leader initiation site. The TATA box is thought essential for the activity of DNA-dependent RNA polymerase involved in transcription. There is also a highly conserved octamer with the concensus sequence ATGCAAAT (also present in the kappa chain promoter) lying between 90 and 150bp upstream of the leader initiation site which is critical for transcription to proceed. However, these elements need to interact with enhancer sequences which are located between the most downstream J segment and C_μ in heavy chain genes or the analogous intron in light chain genes. The enhancer is so positioned 5' of the newly expressed constant cluster thereby exerting its effects. The interaction can only occur when rearrangement has taken place because the distance between the promoter and enhancer elements is then significantly reduced. Hence transcription does not occur in non-rearranged germline genes. These elements are cis-acting because they control the transcription rates of their cognate genes.

transacting
nuclear factors

This cis-interaction of promoter and enhancer elements requires yet another element to promote transcription in B cells - the trans-acting nuclear factors. One of these NF-κB is found in the cytoplasm of B cells complexed to another protein I-κB. When transcription is about to start the NF-κB dissociates from the I-κB and moves to the nucleus where it binds to a 10bp sequence in the enhancer. That is really all we know about the process at the moment! The details are summarised in Figure 3.15.

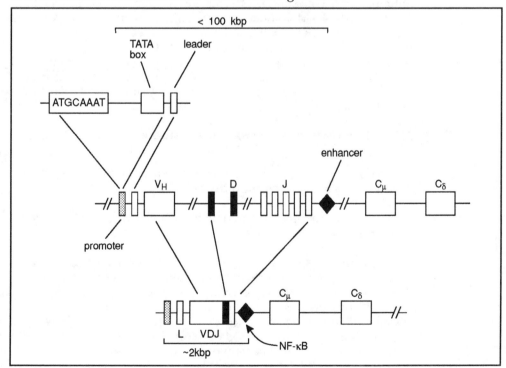

Figure 3.15 Regulatory elements in immunoglobulin gene transcription.

3.12 Molecular genetic approaches to the humanising of antibodies for therapy

Having come this far, you now have a considerable knowledge of immunoglobulin genes and antibodies. It has been realised for some time, especially with the advent of monoclonal antibody technology, that antibodies have a great future as immunotherapeutic agents if we can overcome one or two major problems. We have not been very successful in producing human monoclonal antibodies due firstly, to ethical and technical problems in obtaining suitable B cells to the antigen of interest and secondly, to difficulties in finding fusion cell partners for human B cells. So rats and mice are still the main source of such antibodies.

ethical and technical problems

chimaeric antibodies

However, as you will have learnt from the last section of Chapter 2, these xenogenic antibodies are immunogenic in Man and this presents major problems for therapy. However, researchers have attempted to combat this problem by engineering hybrid antibodies, often called chimaeric antibodies, in which the most immunogenic regions, the Fc portion, of the rat or mouse antibodies had been replaced by human sequences. More recently, human antibodies have been engineered in which only the CDR regions are xenogenic. This, of course, has led to the idea that we can generate new specificities or improve the affinity of antibodies by introducing small base changes in the DNA sequences of any of the CDR's of heavy or light chain genes. We shall now have a brief look at how these antibodies are engineered.

This section assumes that you have a basic knowledge of DNA cloning including the use of restriction endonucleases, cloning vectors and transformation of *E. coli* using antibiotic resistance markers (This knowledge can be gained in outline from the BIOTOL text 'The Molecular Fabric of Cells' or in more detail from the text 'Techniques for Engineering Genes'). If you have not studied gene manipulation then you might find the details of the remaining few pages of this chapter difficult to follow. The point is that we are explaining that by employing genetic engineering techniques, we can, at least in principle produce antibodies which have improved properties (eg greater affinity, greater immunological tolerance). If you have not studied gene manipulation do not, at this stage, attempt the final SAQ.

3.12.1 Production of chimaeric antibodies

The first experiments demonstrated that selected cloned immunoglobulin genes inserted into suitable vectors could be introduced into plasmacytoma cells resulting in production of antigen specific IgM antibodies.

Let us examine the special requirements of a plasmid vector to be used to transfect and be expressed in mammalian cells.

Firstly, it must have a prokaryotic origin for bacterial replication and an antibiotic resistance marker so that the recombinants can be replicated before they are used to transfect mammalian cells. Secondly, the vector will need eukaryotic elements which control transcription initiation in mammalian cells. Thirdly, it will need polyadenylation elements and, perhaps, sequences which promote processing of transcripts. Lastly, some method of selection of the transformants after transfection of the mammalian cells will be required.

pSV2
prototype
vector

Most of the experiments involving Ig genetic engineering and expression use the pSV2 prototype vector introduced by Mulligan and Berg in the early 1980's. This has an ampicillin resistance gene and an origin of replication from the bacterial plasmid pBr322. These will be useful in the replication and cloning of the recombinants. pSV2 also has promoters and polyadenylation elements from simian virus 40 (SV40) which fulfil the second and third requirements above. Derivatives of pSV2 such as pSV2gpt and pSV2neo also have resistance genes to drugs such as mycophenolic acid or kanamycin respectively. EcoRIgpt in pSV2gpt codes for the enzyme hypoxanthine phosphoribosyl transferase. This enzyme is absent in some myeloma cell lines and this can be the basis of the test for transformants of myeloma cells transfected with recombinant pSV2gpt vectors. The principal features of pSV2 are shown in Figure 3.16.

pSV2gpt

transformants
of myeloma
cells

Using such a vector, the way was open to examine the feasibility of producing functional chimaeric antibodies composed of murine variable domains and human constant domains. There are three main stages in the procedure.

shuffle the
exons

Firstly the exons have to be shuffled. This means that using restriction sites common to both the vector and Ig gene fragments the murine rearranged variable gene can be introduced together with the human constant gene into the same vector. In a similar fashion the murine variable light chain gene can be transferred together with the human light chain constant gene into a vector. The particular restriction fragments and vector restriction sites are shown in Figure 3.16. Notice that it is possible to split the genes due to the presence of a restriction site situated between the variable and constant genes in each case. Examine Figure 3.16 carefully as it shows the various stages involved in producing chimaeric antibodies.

The second stages involves the propagation of the vector in *E. coli* using the ampicillin resistance marker to identify transformants and recover the pSV2 vectors.

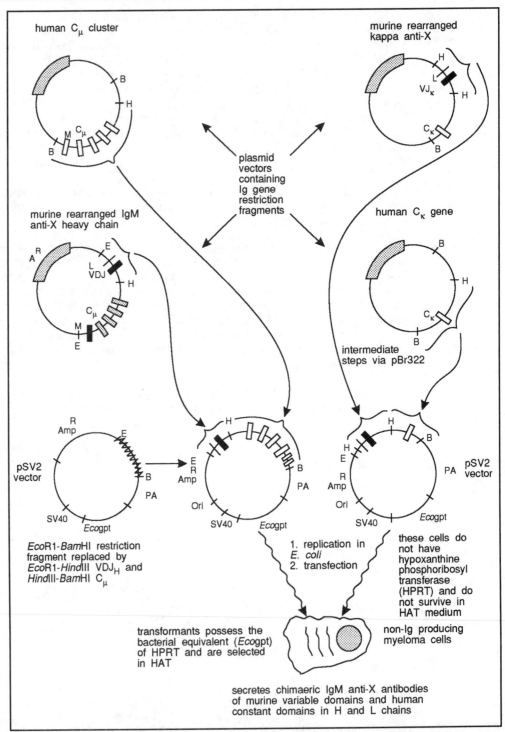

Figure 3.16 Production of chimaeric antibodies E = *Eco*R1 restriction site, B = *Bam*H1 restriction site, H = *Hind*III restriction site, Ori = replication organ, PA = polyadenylation gene, Amp[R] = ampicillin resistance, A[R] = antibiotic resistance (gene used for selection of recombinants in bacteria), M = membrane exons. HAT medium = culture medium containing hypoxanthine, aminopterin and thymidine. Note that transformed cells can grow in HAT medium, untransformed cells cannot.

transfection

The final stage involves transfection of the myeloma cells using a variety of methods and looking for the expression of the chimaeric genes. The myeloma cells used in most experiments were SP2/0 in which the Ig genes are not expressed. These cells do not make the enzyme HPRT and cannot survive in medium containing aminopterin. However, expression of pSV2gpt in the transformants results in the production of HPRT and hence transformants survive. Some of the transformants produced functional chimaeric antibodies. The full procedure is summarised in Figure 3.16.

This ability to produce mouse-human chimaeric antibodies is a significant advance towards the preparation of immunotherapeutic agents since the presence of the human heavy and light chain isotype should significantly reduce the immunogenicity of the antibodies. Such approaches also provide us with the facility of choosing the isotype or subclass possessing particular effector functions. Also, the use of the murine variable regions avoids the difficulties encountered in obtaining human antibodies of the desired specificity. However, we can do better than this as will be seen below.

3.12.2 Reshaping antibodies

oligonucleotide-
directed
mutagenesis

Oligonucleotide-directed mutagenesis is a procedure whereby you can introduce changes or mutations in DNA coding for the antibody. This is usually in the form of copy (complementary DNA) DNA (cDNA) made from using the mRNA for the antibody and the enzyme reverse transcriptase. The gene to be altered has to be in single stranded form so the cDNA strand is ligated into a M13 vector (a single stranded DNA vector). A short oligonucleotide is then synthesised that is complementary to the DNA except at the mutation point. The oligonucleotide is allowed to anneal (form a double stranded DNA molecule) to the cDNA strand and acts as a primer for DNA polymerase 1 (Klenow fragment) that produces the complementary strand containing the altered sequence. On transfection into *E. coli* and DNA replication, half the phage particles will contain the wild type double stranded DNA (unmutated) and the other half will contain double stranded DNA where both strands contain the altered sequence. Phage plaques produced on the agar plates can be examined for the new sequence by hybridisation probing using the radiolabelled oligonucleotide as the probe.

primer for DNA
polymerase 1

immunotherapy

reverse
transcriptase

Using this technique, various groups have increased the affinities of antibodies by amino acid changes in the CDRs and sometimes even in framework regions while others have replaced the CDRs of human with those of mouse. This approach obviously has great potential in immunotherapy as it can avoid the necessity of making human monoclonals of a particular specificity. Instead, you make a monoclonal antibody of the desired specificity, extract the mRNA from the hybrids, prepare the cDNA using reverse transcriptase and then clone and sequence it. Using oligonucleotide-directed mutagenesis we could then exchange the murine CDR sequences for the human thus producing the mutant human cDNA. This could then be ligated into a suitable vector such as pSV2 and transfected into non-Ig producing myeloma cells as we discussed earlier. The techniques are not yet as straight forward as we may appear to suggest but we can expect to see rapid progress towards actual immunotherapeutic goals in the near future.

SAQ 3.11

Examine the 2 vectors pBr322 and pSV2neo and decide whether you could successfully ligate the murine light and heavy variable genes and the human light and heavy constant genes into the vectors using only the restriction sites shown in Figure 3.17 for transfection into SP2/0 for production of functional IgM antibodies consisting of murine variable domains and human constant domains. Which one of the following statements is correct?

1) The murine IgM variable restriction fragments could not be inserted into pSV2neo along with the human IgM constant gene restriction fragment.

2) The restriction fragments for the chimaeric light chain can be ligated into pBr322 and this can transfect SP2/0.

3) The chimaeric light chain restriction fragments can be ligated into pSV2neo.

4) Although the *EcoRI-Hind*III fragment (L-VDJ) of the murine heavy chain gene and the *Hind*III-*Bam*HI fragment (C$_\mu$) of the human IgM heavy chain gene can be successfully ligated into pSV2neo, this cannot be done with the light chain restriction fragments. Therefore, only the heavy chain gene would be expressed in SP2/0.

5) The SP2/0 would produce functional IgM antibodies.

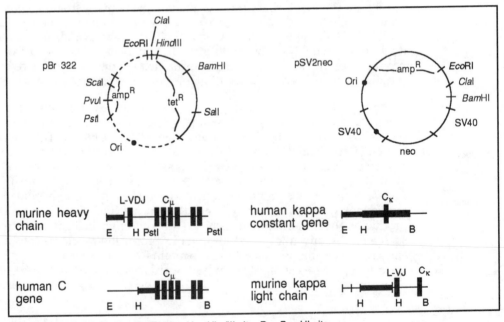

Figure 3.17 SAQ 2.11. E = *Eco*RI site; H = *Hind*III site; B = *Bam*HI site.

Summary and objectives

In this chapter, we have studied in depth the molecular and genetic events associated with antibody production and diversification. We have explained how the genes coding for antibodies are rearranged during the development and maturation of B cells. We have also discussed the molecular mechanisms associated with gene rearrangement and RNA processing. These discussions enabled us to explain how antibody diversity, class switching and the production of soluble and membrane bound antibodies is achieved. This knowledge is facilitating the design and production of antibodies of high therapeutic value.

Now you have completed this chapter you should be able to:

- define the chromosomal arrangement of immunoglobulin light and heavy chain gene segments;

- describe the basic DNA rearrangement mechanism;

- list the features of light and heavy chain genes;

- explain the joining rules for DNA rearrangement and the role of spacers, heptamers and nonamers;

- discuss the contribution made by junctional diversity, N regions, somatic mutation and V_H exchange to antibody diversity;

- discuss immunoglobulin germline gene expression in CD5$^+$ B cell differentiation and development;

- explain in detail the term immunoglobulin class switching;

- explain how the B cell selects for membrane and surface IgM synthesis;

- formulate procedures for shuffling exons in vectors to produce chimaeric antibodies and introduce mutations using oligonucleotide-directed mutagenesis.

Cell interactions in antibody production

Cell interactions in antibody production

4.1 Landmark experiments on cell interactions in the humoral response

Our current knowledge of cell interactions involved in antibody production derives from experiments by a number of researchers who demonstrated a fundamental role for small lymphocytes in immune responses. We shall begin by describing just a few experimental approaches that gave us the conceptual ideas on cell interactions.

We will go on to explain the features of the major histocompatability complex which plays an important role in cell interactions. Subsequently we will describe the molecular and biological consequences, of these interactions.

4.1.1 The lymphocyte, the thymus and the bursa

You may find this chapter conceptually quite difficult. In order to give you a thorough knowledge of how cells interact in stimulating antibody production you will have to learn a lot about a variety of cells and their surface components. For many reasons a kind of jargon in the form of abbreviations, letters and symbols has developed. We have tried to expose you to this terminology but you might find it a little difficult to remember them all immediately. We remind you however that we have provided a list of abbreviations and their meaning at the end of this text. We have also provided a table summarising the features of the main types of cells encountered in the immune responses. Make use of these aids to help you get to grips with this important section.

immune
deficiency

Gowans, in 1962, showed that if he removed lymphocytes from animals by cannulating the thoracic duct this resulted in a severe immune deficiency state that could be corrected by restoring the lymphocytes to the animal. This demonstrated that these recirculating lymphocytes played a major role in the production of antibodies.

exterpation
experiments

Further progress was made by removing organs from experimental animals, so-called exterpation experiments, and observing the effects on the *in vivo* response. Glick demonstrated in the chicken that the removal of the Bursa of Fabricius, a small organ situated near to the cloaca resulted in an inability to produce antibodies. The removal of the Bursa, however, had no effect on cell mediated immune reactions such as graft

neonatal
thymectomy

rejection. Also in the early 1960's, Miller demonstrated that neonatal thymectomy (ie removal of the thymus at birth) resulted in a loss of both humoral and cell mediated immunity.

4.1.2 The lethally irradiated and thymectomised mouse

A number of landmark experiments were then reported which demonstrated the requirement for B cells, T cells and macrophages for induction of humoral responses and that these cells had to be from genetically related animals for these cells to collaborate with each other.

thymectomy
and lethal
irradiation

reconstituted
mice

These experiments became feasible because of the development of the 'adoptive transfer technique' in which mice were made totally immunoincompetent by thymectomy and lethal irradiation that destroyed the lymphoid cells and the bone marrow stem cells. However, these mice could be made competent by replacement of these cells from untreated genetically identical mice. Such transfers are called isografts or syngrafts and the mice were said to be reconstituted mice. We shall refer to these animals as the recipients, the transferred cells being supplied by donors.

4.1.3 Thymus-bone marrow synergism in the response

An experiment was performed by Claman and coworkers who transferred thymus and/or bone marrow cells into mice rendered immunoincompetent by the treatment described. The recipient mice were then immunised with antigen. In this experiment, sheep red cells (SRC) were used as the antigen, and the spleens of these mice were examined a few days later for the production of antibodies.

donor cells

Since the host lymphoid cells had been totally destroyed, any response observed in this animal would have been due to the interactions between populations of donor cells. In the absence of any such transfer, the recipient could never mount a response as the bone marrow cells had been destroyed; stem cells in the bone marrow are the precursors of all hematopoietic cells including cells of the thymus. So a positive response in the recipient animal would indicate that the donor cells were able to respond to the antigen.

The results demonstrated that bone marrow cells alone were not able to respond to the SRC and thymocytes responded only minimally. However, if both thymocytes and bone marrow cells were transferred, the response which ensued was equivalent to that of splenic lymphocytes. The latter, of course, contains both T and B cells as well as macrophages. This experiment showed that the production of antibodies required at least 2 populations of cells, one from the bone marrow, the other from the thymus. Claman suggested that cells of either the thymus or bone marrow population produced the antibodies whilst the other contained an auxiliary cell that was necessary for antibody production. In the next few years, other workers demonstrated that B cells within the bone marrow population were solely responsible for producing the actual antibodies.

4.1.4 The B cell response requires carrier specific T cells

Do you remember what a hapten and carrier are? Let us summarise the principal features of these molecules. A hapten is antigenic (binds to antibody) but is not immunogenic (cannot induce an antibody response by itself). The hapten must be physically coupled to an immunogenic molecule such as a protein (the carrier) to induce anti-hapten antibodies. To be more specific, DNP (dinitrophenol) is a hapten, used in many of the early experiments, that cannot by itself induce anti-DNP antibodies. However, if we attach it covalently to bovine serum albumin (BSA) and use this as the immunogen, then anti-DNP antibodies will be produced. Anti-BSA antibodies will also be produced since BSA by itself is an immunogen. If you give a second dose of the DNP-BSA to the animal you will observe a secondary immune response. However, if you link the DNP to a different carrier and use this for the second injection you will only observe a primary response. These observations illustrated the 'carrier effect' and suggested that the determinants or epitopes on the carrier and those of the hapten were being 'seen' by different cell types.

carrier effect

This was demonstrated in a series of experiments. In summary, one group of mice were injected with hapten-carrier 1 (H-Cl) and another group with carrier 2 (C2). Spleen cells were then transferred from both groups of immunised (primed) mice into irradiated recipients and the latter injected with various antigens as shown in Figure 4.1. In some experiments, the spleen cells were treated with an anti-thymocyte (we shall call it anti-T) antiserum + complement to eliminate all T cells before completing the transfer.

∏   Before you read on, examine the results in Figure 4.1. What conclusions can you draw from the experiment?

Let us take this step by step. In the first experiment we have transfered cells which have been primed with H-C1 to the irradiated (immunoincompetent) mouse. On challenging this mouse with H-C1 we get a typical secondary anti-hapten (H) response. In the second experiment, when the immunocompetent mouse receive donor cells which have been primed with H-C1, these cells do not respond when challenged by H-C2. In experiment 3, the recipient mouse receives cells which have been primed with H-C1 (from Donor 1) and with cells primed with C2 (Donor 2). When challenged with H-C2, then a secondary anti-H response is produced. This indicates that there is some form of co-operation between the two sets of donor cells. Destroying the T cells from the C2 primed donor with anti-T antiserum, abolishes the secondary response to H-C2.

This indicates that recognition of the carrier part of the challenge is carried out by T cells. The production of antibodies is of course, carried out by B cells. Thus it would appear that the hapten part is recognised by B cells and the carrier part by T cells.

Note that challenge with the wrong conjugate (experiment 5 in Figure 4.1) does not lead to a secondary response. Thus in this experiment Donor 1's cells were primed with C1 and Donor 2's cells with C2. Even working together they could not respond to H-C3. The final experiment indicates that the hapten must be attached to the appropriate carrier in order to induce a secondary anti-H response. Of course there are many combinations in which these experiments could be carried out.

In summary therefore the answer to the intext activity posed about the results shown in Figure 4.1 should be that a) different cell types were recognising the hapten and carrier determinants, b) these cells collaborated if presented with the correct hapten-carrier conjugate and c) the hapten had to be covalently attached to the carrier to induce anti-hapten antibodies.

antigen bridge   These results suggested that there was an antigen bridge consisting of hapten and carrier linking the B cell and T cell together. The workers at that time envisaged that the B cell antibody receptor bound the hapten determinant and the T cell receptor bound a determinant on the carrier. In this way the cells were brought close together for delivery of activation signal(s) between the cells. We shall learn of a different interpretation later.

a) design of the experiments

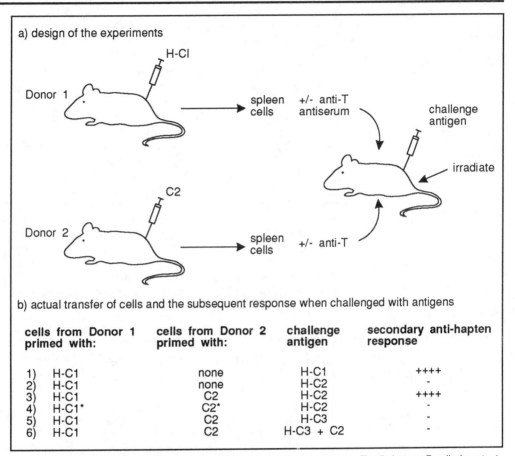

b) actual transfer of cells and the subsequent response when challenged with antigens

cells from Donor 1 primed with:	cells from Donor 2 primed with:	challenge antigen	secondary anti-hapten response
1) H-C1	none	H-C1	++++
2) H-C1	none	H-C2	-
3) H-C1	C2	H-C2	++++
4) H-C1*	C2*	H-C2	-
5) H-C1	C2	H-C3	-
6) H-C1	C2	H-C3 + C2	-

Figure 4.1 The carrier effect. * Treated with anti-T antiserum which destroys T cells but not B cells (see text for a discussion). Note that in these experiments only anti-hapten antibodies were assayed. Experiments in b) are numbered on the left hand side.

SAQ 4.1	Which one of the following statements is correct with reference to the experiments described in Figure 4.1.

1) In the mice primed with hapten-carrier 1, B cells recognise only hapten and T cells the carrier.

2) Hapten-carrier 3 did not induce any response in the recipient mice.

3) Spleen cells removed from recipient mice given hapten-carrier 3 + carrier 2 and transferred to another irradiated mouse would produce a secondary anti-hapten response to hapten-carrier 3.

4) Spleen cells from mice primed with hapten-carrier 1 and transferred with anti-T antiserum-treated spleen cells from carrier 2 primed animals would give a secondary anti-hapten response in recipient mice injected with hapten-carrier 2.

5) Treatment of spleen cells from hapten-carrier 1 primed mice with anti-T antiserum prevented the secondary response to hapten-carrier 2 in recipient mice when transferred along with carrier 2 primed T cells.

4.1.5 Macrophages are required to prime T cells

priming of T
helper cells

Experiments in the mid-1970s demonstrated a requirement for macrophages during the priming of T helper cells in antibody responses. One such experiment was performed as follows.

An enriched population of macrophages was obtained from the peritoneal cavity by injection of phosphate buffered saline and collection of the cell suspension with a syringe a short time later. The macrophages were then treated with anti-T antiserum + complement to remove T cells and these were then mixed with T cells from lymph node suspensions from which B cells and macrophages had been removed (see Figure 4.2). The macrophage and T cell populations so obtained were only 'enriched' for either cell type and were not pure populations. The cells were mixed together with antigen to prime the T cells in a culture dish (*in vitro*). These primed T cells were then able to provide help to B cells in a subsequent *in vitro* response to antigen. If the macrophages were absent in the priming procedure, the T cells were not able to support the B cell response.

antigen
presenting cells

dendritic cell

This experiment suggested a role for macrophages in the activation of unprimed T helper cells. As we shall see later, cells which activate T helper cells are called antigen presenting cells (APCs). These include only some types of macrophages, B cells and other cells as well. In fact, current thinking assigns the role of T cell activation to APCs called interdigitating dendritic cells or immunostimulatory cells found in sites such as the lymph node paracortex. B cells cannot activate unprimed T cells but are potent APCs in their own right once the T cell has been primed. However, after such experiments as this, it was assumed that macrophages were the principle cells that activated T lymphocytes - a justifiable conclusion under the circumstances. We shall maintain this assumption for the moment.

∏ Explain why anti-Ig + C′ when added to lymph node cells destroy B cells (see Figure 4.2).

You should have realised that anti-Ig would bind to antibodies on the surface of the B cells. Subsequently the binding of complement proteins leads to the lysis of these cells.

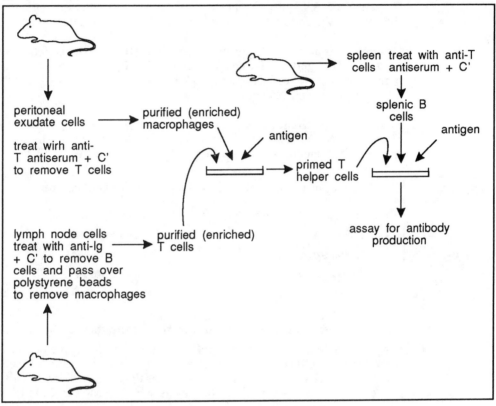

Figure 4.2 Evidence for a macrophage requirement in priming of T cells.

4.2 The Major Histocompatibility Complex

We are about to examine the role of the MHC in cell interactions. However, before we do that it is necessary to study in more depth the Major Histocompatibility Complex that you studied in Chapter 1. We remind you of some important facts.

- Cytotoxic T cells recognition of virally infected cells requires viral antigens to be present on the surface on the infected cell. The cytotoxic T cells recognises both the viral antigen and the MHC Class I components on the surface.

- T helper cells recognise MHC Class II products on B cells. They are only capable of activating B cells they recognise.

We also remind you of a few terms used in genetics.

Common genetic terms

Locus This is the position of a single gene on the chromosome.

Polymorphism Most of our genes on both maternal and paternal chromosome have a single nucleic acid sequence encoding a single protein of which there are no variants. In a few genes, of which the MHC genes are an excellent example, the nucleic acid sequence varies slightly so that a number of variants of the protein are produced within *allele* the species. We call such genes polymorphic and each gene variant is called an allele.

Homozygous and heterozygous If an individual possesses different alleles at a genetic locus on the maternal and paternal chromosomes we say that individual is heterozygous; if identical alleles are present at the genetic locus on both chromosomes, the individual is homozygous.

Syngeneic and allogeneic If two individuals are genetically identical such as identical twins or mice of the same inbred strain, they are syngeneic; all other members of the species are allogeneic to one another ie they express different alleles.

SAQ 4.2

Match the items on the left with those of the right column.

1) Antigen presenting cells   a) Same alleles

2) Polymorphism   b) Self recognition

3) Homozygous   c) T cell priming

4) Allogeneic   d) Many alleles

5) MHC   e) Non-identical, same species

4.2.1 Congenic mouse strains and the discovery of the MHC

syngrafts

isografts

allografts

Early observations showed that grafts of tissue between syngeneic animals (syngrafts) or to another site on the same animal (isografts) were not rejected whereas those between animals of different strains (allografts) were rejected. This is the result of recognition of 'self' or 'nonself' tissues as we discussed in Chapter 1 and the genes responsible for this self/nonself recognition were called histocompatibility genes.

congenic strains, Major Histocompatibility Complex, H-2, HLA

It is possible to construct by a complex process of inbreeding, strains of mice which only differ in the genes causing transplant rejection. All the other genes of the two strains are identical - we say the mice have the same background but a different MHC. These mice are called congenic strains and using these mice it was discovered that the MHC consisted of a complex of different but closely linked genes responsible for graft rejection (Class I genes). This region was named the Major Histocompatibility Complex, H-2 in mouse and HLA in Man.

Subsequently, it was discovered that this region also contained genes which controlled T helper cell activity (Class II genes) and some genes which encoded some complement proteins (Class III). The latter have nothing to do with cell interactions; it just happens that the term MHC included the chromosomal segments that contained these loci. As you will see later, there are other non-cell interaction genes present in this complex also.

Π   You may want to refer back to Chapter 1 (Section 1.5) to remind you of the roles of MHC Class I and II.

4.2.2 The genomic organisation of the MHC

The MHC is situated on the short arm of chromosome 6 in Man and chromosome 17 in mouse and Figures 4.3 and 4.4 present simplified versions of the MHC in Man and mouse respectively.

∏ In the next few pages we are going to provide you with a lot of information about the genome organisation of the MHC. You may need to to read it through a few times to really remember it all. You will find Figures 4.3 and 4.4 useful summaries. It would be good practice to read through this section and then draw your own versions of Figures 4.3 and 4.4. Eventually you will be able to do this without referring to the text or Figures 4.3 and 4.4. You will then be able to truly claim to know how the MHC is organised.

Class II genes

Starting from the centromeric end of the chromosome, we find a cluster of Class II gene loci in Man called DP, DQ and DR. Each of these code for proteins called α and β chains which are expressed in non-covalent association on the cell surface as pairs. Any of these loci may encode more than one α and β chain and any α chain can pair up with any of the β chains encoded within the same locus. For instance, the HLA-DR locus has one α chain gene and 3-4 β chain genes. Thus up to 4 different $\alpha\beta$ pairs can be expressed on the surface of any one cell. If the individual is heterozygous this number increases to 8 for the HLA-DR locus alone! However, some of the genes are non-functional (pseudogenes) or the protein product has not yet been identified so the actual number of possible gene products is somewhat less than this. As you can see DP and DQ have lower numbers of genes. It has been calculated that an individual may express up to 20 different Class II gene products on the surface of any one cell. In addition, there are other ill-defined Class II-like genes called DZ, DO and DX. The DP, DQ and DZ genes are expressed on antigen presenting cells and on activated helper T cells.

The murine MHC Class II consists of the I-A and I-E subregions which, like Man, encode α and β chains which are expressed on the cell surface as $\alpha\beta$ pairs. The I-A region codes for the Aα and Aβ chains and also a highly polymorphic Eβ chain whereas the I-E region encodes the Eα chain. As in Man, there are a number of β chain genes in both I-A and I-E (some of which are pseudogenes) that suggests multiple $\alpha\beta$ pairs can be expressed on the cell surface.

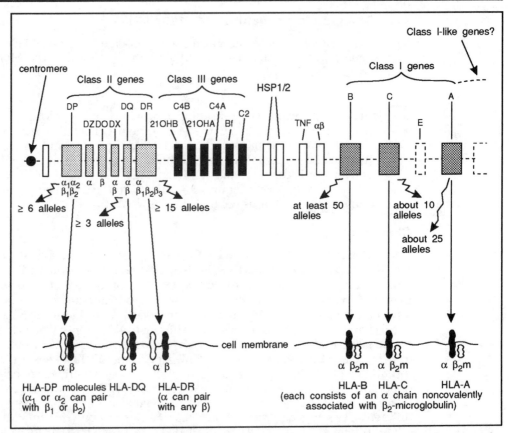

Figure 4.3 The human HLA region and the gene products (not to scale). Note that the Class III genes code for complement proteins and serum factors (further details are given in the text and Figure 4.4). TNF$_{\alpha\beta}$ are tumour necrosis factors. 21 OHA and 21 OHB are genes coding for steroid biosynthetic enzymes, 21-hydroxylase A and B. HSP = heat shock protein genes. These are also recorded in the text.

Class III genes

The Class III genes are situated next to the Class II cluster in Man and these encode complement components C2 and C4, serum factor B of the alternate complement pathway and the Slp (sex limited protein) which is a non-functional variant of C4 that can be expressed under the influence of testosterone. These products are actually secreted by the cell. This cluster also includes genes for 2 enzymes involved in steroid biosynthesis, 21-hydroxylase A and B. As we have emphasised previously, there is no known relationship between these genes and the function of the Class I and II genes. They are, however, like the Class I and II genes, highly polymorphic within the species.

Class I genes

At the telomeric end of the MHC in Man we find the Class I genes. They are separated from the Class III genes by loci for genes encoding heat shock protein (HSP) and the cytokines tumour necrosis factor TNF$_\alpha$ and TNF$_\beta$ (also called lymphotoxin).

There are three loci in the order B, C and A. These genes each code for a single α chain which is expressed at the cell surface always associated with β_2-microglobulin which is itself encoded by a gene on chromosome 15 in Man and 2 in mouse.

There are additional ill-defined Class I genes situated mainly telomeric to the HLA-A, some of which are pseudogenes and others which are non-polymorphic but which may be found on the cell surface associated with β_2-microglobulin. They are not thought to be involved in the MHC restriction of cytotoxic T cells (ie in graft rejection).

Murine MHC Class I genes are found in two loci K (K, K_1) and D/L (usually referred to as just D) separated by Class II, Class III and TNF genes. Telomeric to D/L is a large segment of DNA containing Class I-like genes called Qa and Tla (thymic leukaemia antigen). The gene products of these genes are found principally on lymphoid cells in contrast to the classical H2-K and H2-D molecules which are found on all nucleated cells. They are also present in much lower numbers per cell than the K/D molecules and the degree of polymorphism is much less. Although they have not been proved to be involved in Class I restriction, congenic mouse strains differing only at these loci can reject grafts. They are likely to be analogous to the non-A, B and C human Class I genes which as yet are not well defined.

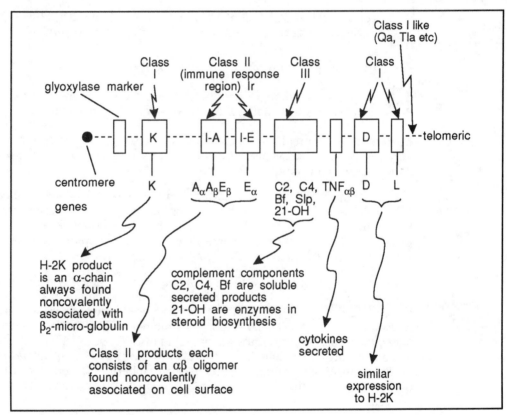

Figure 4.4 Simplified version of mouse H-2 (not to scale).

Note that the nomenclature of the HLA system is internationally fixed. The loci are indicated by means of characters (A, B, C, DR, DQ, DP) and the various alleles are indicated by means of numbers (for example A2, B5 etc) providing there is international consensus as to the antigenic definition of such an allele. If no such international agreement is reached then the number is preceded by a 'w' (for example Bw35, DRw9).

SAQ 4.3	Complete the following sentences using the word list provided:

1) At the [   ] end of the chromosome in Man we find a cluster of genes called [   ], [   ] and [   ]. These are expressed on [   ] cells and activate [   ] T cells.

2) In mice, the [   ] and [   ] loci regulate [   ] T cells. They also induce [   ] when tissues are transplanted into [   ] mice.

3) [   ] genes in Man are found at the [   ] end of the MHC complex. Because of the many [   ] at single [   ], we say they are very [   ].

Word List

centromeric, DP, DQ, DR, cytotoxic, Class I, telomeric, alleles, graft rejection, antigen presenting, allogeneic, loci, helper, polymorphic, K, D.

4.2.3 Polymorphism and inheritance of the MHC

Π Before you read on, re-examine Figure 4.3. Can you make any conclusions as to why the majority of individuals are allogeneic to each other?

By examining Figure 4.3 you should be able to conclude that each individual (except identical twins) is allogeneic ie possesses a combination of alleles for Class I and II molecules which is unique to that individual.

MHC polymorphism

haplotype

alloantigens

Because of the close linkage of the MHC genes (the short distance spanned by the whole MHC), the whole set of alleles at each locus on a single chromosome is normally carried to the progeny. This set is called a haplotype. Since an MHC set is carried on both paternal and maternal chromosomes, we inherit two haplotypes. The allelic forms of Class I molecules are also called alloantigens.

MHC genes are codominant

For MHC Class I then you would inherit one HLA-A, one HLA-B and one HLA-C allele from your father and a similar number from your mother - a total of 6 alleles. Since all these MHC genes are co-dominant these give rise to the expression of 6 different HLA molecules on each nucleated cell. If your mother and father are homozygous for any allele, then this would reduce the numbers of different allelic products on each cell. Since there are more than 50 HLA-B alleles, about 25 HLA-A alleles and 10 HLA-C alleles in the human population, the chances of inheriting exactly the same combination of 6 alleles as another individual are minimal! You can argue similarly for Class II molecules.

linkage disequilibrium

The frequency of finding two alleles at different loci in the same individual should be the product of the frequencies of each of the two alleles in the population. It has been found, however, that some combinations occur at a far higher frequency than expected. This phenomenon is called linkage disequilibrium. Examples of this are HLA-B8 with HLA-A1 and HLA-B7 with HLA-DR2.

4.2.4 HLA-related diseases

HLA-B27 and
ankylosing
spondylitis

Possession of certain HLA alleles appears to predispose some individuals to contracting certain diseases. The most dramatic evidence for this phenomenon is ankylosing spondylitis, an inflammatory disease resulting in stiffening of the spine. Over 90% of patients with this disease carries HLA-B27 allele. You should note that all people with this allele do not contract the disease - possession simply carries an increased risk. Other HLA-related diseases (HLA in parentheses) include Hashimoto's thyroiditis (HLA-DR5), myasthenia gravis (HLA-B8), multiple sclerosis (HLA-DR2) and dermititis herpetiformis (HLA-DR3).

molecular
mimicry

The reason for such HLA connections is not yet clear. It is possible that the particular alleles act as receptors for the pathogens to gain entry to cells. Another suggestion is that the MHC allelic product and the pathogen have high homology which results either in non-recognition of the pathogen as nonself or induction of a reaction against the self tissue (autoimmune reaction). This resemblance is called molecular mimicry.

SAQ 4.4

Examine Figure 4.5. Which of the following is/are correct?

1) Paternal cytotoxic T cells would not recognise virally infected cells of the child as self nor destroy these cells.

2) Paternal cells injected into the child would induce a graft versus host reaction.

3) Maternal cells injected into the child would be destroyed.

4) T helper cells of the child could not promote activation of paternal B cells.

5) Maternal cells injected into the child would not be destroyed because they share a common haplotype.

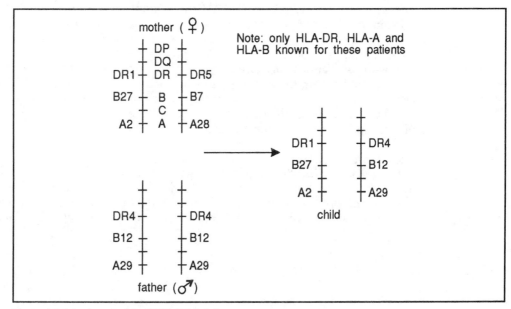

Figure 4.5 Inheritance of the MHC (SAQ 4.4).

4.2.5 Structure of MHC molecules

MHC Class I molecules

These structures are found on all nucleated cells and platelets and consist of single polypeptide chains of about 44kD. The 80% of the structure which lies outside the cell membrane can be divided into three domains of about 90 amino acids in length, α_1, α_2 and α_3 (see Figure 4.6a). Both α_2 and α_3 have a disulphide loop in their domains. There is a hydrophobic transmembrane portion of 25 amino acids and a cytoplasmic portion of about 30 amino acids. The α chain is always found non-covalently associated with the 12kD β_2-microglobulin. This polypeptide interacts with all three external domains, presumably stabilising the structure. Both the α_3 domain and β_2-microglobulin show

immunoglobulin
superfamily

considerable homology with immunoglobulin constant domains and are part of the immunoglobulin superfamily.

We introduced you to the idea that antigens are found associated with MHC molecules on the surfaces of various cells and these complexes activate various T cells. As we shall find later, these are not whole antigenic molecules but short peptides that are found bound to the MHC molecules. The allelic differences (up to about 20% of the amino acid sequence) between different MHC Class I molecules eg between HLA-B8 and HLA-B27 are mainly found in the amino acid sequences of the α_1, α_2 domains and are clustered within a cleft formed by parts of both domains in which can be accommodated peptides of up to 20 amino acids in length. You can imagine then that different MHC Class I molecules will be able to bind different peptides and this is the reason for the polymorphic nature of these molecules. We shall see later that other polymorphic regions of the MHC also interact with the T cell receptor. The structural features of Class I molecules are shown in Figure 4.6a.

MHC Class II molecules

These molecules are constitutively expressed (always found) on a very restricted spectrum of cells including B cells, interdigitating dendritic cells and thymic epithelial cells and are composed of two non-covalently associated polypeptide chains α and β. The α chain of about 34kD and the β chain of about 29kD each possess two extracellular domains α_1 α_2 and β_1 β_2 respectively. Each domain is about 90 amino acids in length and all except the α_1, domain possess disulphide loops of about 60 amino acids. Although we have not yet obtained a crystallographic picture of the peptide binding site analogous to that of Class I we know that the polymorphic regions of both chains are concentrated within the α_1 and β_1 domains and these appear to be folded in a similar fashion to that of the domains within the Class I binding site. Once again, then, we can assume that various alleles of MHC Class II genes facilitate the binding of many different peptides for presentation to T cells.

The α_2 and β_2 domains are also members of the immunoglobulin superfamily bearing structural similarities to immunoglobulin constant domains. Although there are some minor differences between allelic forms, the Class II α_2 and β_2 domains are essentially non-polymorphic (monomorphic) in individuals throughout the species. This is of importance as we shall see later. You will find a diagram of an MHC Class II molecule in Figure 4.6b.

Ⅱ Redraw Figures 4.6a and b) but this time place them, side by side. You should be able to make a list of their similarities. For example, look at the number and position of the various domains, the position of the peptide binding cleft, papain cleavage site, the presence of transmembrane regions. In redrawing these you will also become more conscious of their differences.

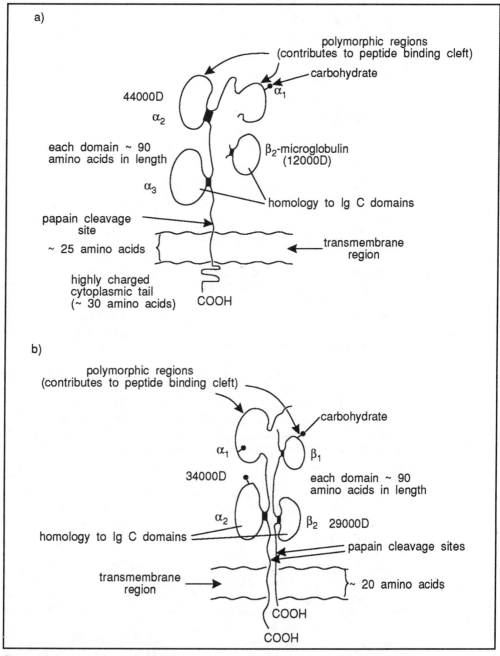

Figure 4.6 Structural features of MHC Class I and II molecules. a) Class I molecules, b) Class II molecules. ■ disulphide bonds.

SAQ 4.5

Complete the following table indicating your choice by + or -.

Characteristic	Class I	Class II	Class III
Present on all nucleated cells			
3 extracellular domains			
Chains of 44kD			
β_2-microglobulin			
Innate immunity connection			
Stimulate helper T cells			
Bind peptides			
Target for cytotoxic T cells			
Polymorphism			
$\alpha \beta$ gene product			
Soluble products			

4.3 MHC and mouse strains

haplotypes and mouse strains

We need to explain some conventions. If you look back to Figure 4.4 you will see that each mouse express K, A, E and D genes. Obviously in a mixed population of mice there will be a variety of combinations of these alleles. To simplify description particular collections of allele combinations of reference mouse strains is given a lower case letter. For instance one mouse strain C57BL/10 is given the haplotype H-2^b and each of the alleles in its H-2 on both chromosomes (it is homozygous) is designated as being b. Similarly, the haplotype of BALB/c mice is H-2^d and DBA mice are also H-2^d. This means that cells of these two strains would recognise each others' cells as self. Helper T cells from a BALB/c mouse would help B cells from a DBA mouse. However, no such help would be forthcoming for B cells from C57BL/10 mice since their haplotype is allogeneic to those of BALB/c and DBA mice. Notice that the mouse strains with the same MHC haplotype are not identical. Although having identical MHC, they have a different background, hence the different strain name.

∏ Before you read on, examine Table 4.1. You should not worry about the actual letters identifying the mouse strain. F1 strands for the first generation after a cross. All except the F1 generation shown are homozygous with respect to H-2. CBA/B10 designates that CBA have been crossed with B10 and then inbred. Can you identify the congenic strain?

Table 4.1 gives you some examples of mouse strains for your reference in the later text. Notice that F1 strains carry both alleles and these are expressed on the cells since the MHC genes are co-dominant. The CBA/B10 is the congenic strain since it carries CBA background genes (looks identical to CBA mice) but carries a homozygous pair of H-2^b derived from B10 mice instead of the H-2^k of the CBA strain. We could in this case regard CBA as the background strain and strain B10 the MHC donor.

Mouse strain	Haplotype	K	A	A	E	E	D
C57BL/10 (B10)	H-2^b	b	b	b	b	b	b
CBA	H-2^k	k	k	k	k	k	k
(CBA x C57BL/10)F1	H-2$^{k/b}$	k/b	k/b	k/b	k/b	k/b	k/b
CBA/B10	H-2^b	b	b	b	b	b	b

Table 4.1 H-2 haplotypes of some mouse strains.

Now let us return to our study of cell interactions!

4.4 T and B cell interactions

4.4.1 T cell-B cell collaboration is MHC Class II restricted

A typical experiment to investigate T and B cell collaboration is shown in Figure 4.7. This experiment shows the result of adoptive transfer of B cells and T cells from different strains of mice primed with hapten carrier (for primed B cells) or heterologous carriers (for primed T cells) to F1 mice followed by priming with hapten on a heterologous carrier as in the earlier experiment.

∏ Examine Figure 4.7 carefully and read its legend. Then decide what conclusions can be drawn from this type of experiment.

Careful examination of the data should reveal that T cells and B cells can only co-operate providing they share haplotypes. Thus it was found that co-operation between B and T cells from mouse strains of unrelated background but identical MHC (congenic strains) was good, whereas cells from mice of identical backgrounds but different MHC did not co-operate. In other words, the MHC Class II molecules on the B cells had to be recognised as self by the T helper cells.

Other experiments led to a similar conclusion for the T cell-macrophage interaction. It was concluded that all APC-T helper cell interactions are governed by the requirement that the T cell must recognise self MHC Class II to provide help. Subsequently, it was found, using recombinant strains, that co-operation between APCs and T cells was dependent on similarities in the I (I-A, I-E) region of the mouse H-2 complex. It was also

MHC Class II restricted

shown that APC - T helper cells interaction was determined by the MHC Class II region (HLA) in Man. We say that the T helper cells are MHC Class II restricted.

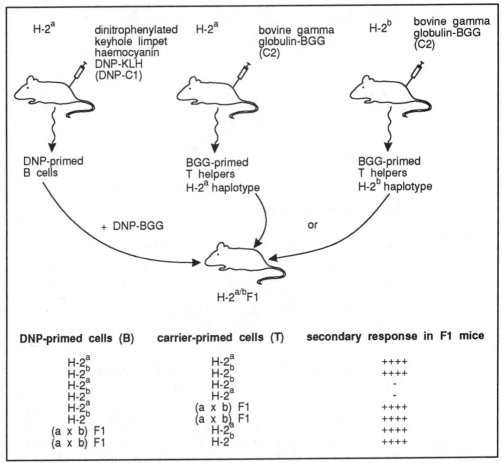

Figure 4.7 Demonstration of MHC restriction in T-B cell interactions. At the top of the figure we show the general design of the experiment. One set of mice are primed with DNP-linked to carrier 1 (C1) and their B cells collected. These are then administered together with T helper cells from other mice which have been primed with carrier 2 (C2) into an immuno-incompetant mouse. This mouse was then injected with DNP-linked to carrier 2 and the immune response recorded. In the lower part of the figure we show the various combinations of haplotypes which were used and whether or not a secondary response (that is a rapid rise in anti-DNP antibodies) occurred in the adaptive mouse. Note that (a x b) F1 will produce H-2$^{a/b}$ halotypes.

4.4.2 Cytotoxic T cells are MHC Class I restricted

An important piece of evidence for MHC restriction came from experiments reported by Zinkernagel and Docherty on assays of virus specific cytotoxic T cell (CTL)-mediated killing of lymphocytic choriomeningitis virus infected human and murine target cells. These target cells were only lysed if they expressed MHC Class I molecules recognised as self by the CTL. So we say that cytotoxic T cells are MHC Class I restricted.

4.4.3 Distinct populations of T cells exist for each parental haplotype

The question then arose as to whether the T cell could recognise one or both the MHC Class molecules expressed on cells of the F1. A number of workers demonstrated by various approaches that there were two distinct populations of T cells in F1 mice and provided evidence that a single T cell can only recognise one MHC haplotype and not both.

Let us examine the experiment shown in Figure 4.8. B6, DBA and (B6 x DBA) F1 mice were immunised with the carrier KLH (keyhole limpet haemocyanin) and T cell-free peritoneal exudate cells (see Figure 4.2) were obtained from each mouse strain. It had been previously shown that macrophages could take up antigen and express it on their surfaces and this in some way activated T cells. The antigen-pulsed macrophages were then injected into (B6 x DBA) F1 mice and enriched splenic T cells were recovered from the mice and examined for their ability to provide help to B cells expressing both parental haplotypes (F1) or a single parental haplotype B6 or DBA.

∏ Before you read on, examine Figure 4.8. Can you explain the results?

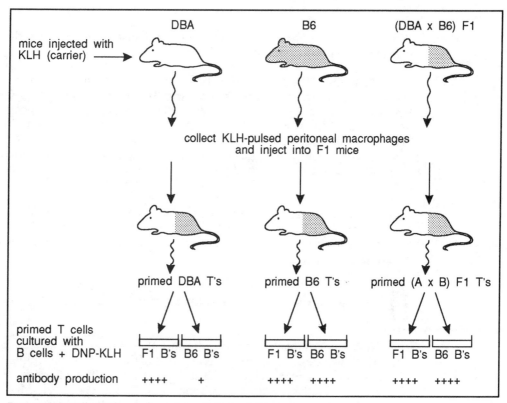

Figure 4.8 T helper cells only recognise a single haplotype.

How did you go on? Let us briefly examine these results. The antigen-pulsed macrophages from DBA mice express DBA molecules on their surface, B6 macrophages express B6 molecules and F1 macrophages co-dominantly express both DBA and B6 on their surfaces. You will notice that B6 B cells received help from F1 T cells that had been exposed to antigen-pulsed macrophages from F1 or B6 mice but very little help from F1 T cells primed by antigen-pulsed macrophages from DBA mice. We have not shown it in the figure, but the converse is also true. Thus DBA B cells received help from F1 T cells primed by F1 or DBA macrophages but not by B6 macrophages. Even though the F1 T cells expressed both DBA and B6 molecules on their surfaces, it appears that single primed T cells only provide help to B cells of one parental haplotype.

Along with results from other experiments, it was concluded that there are separate T cell populations that recognise each MHC II molecule expressed on antigen presenting cells as self. It is also important to notice that both DBA and B6-restricted T cells were specific for the same antigen, KLH. Similar rules govern the recognition of Class I molecules by cytotoxic T cells.

4.4.4 *In vivo* **selection of T helper cells is MHC-restricted**

In support of these findings on MHC restriction, it was discovered that T helper cells recognised antigen + MHC *in vivo*. Earlier findings by Gowans and coworkers had showed that T cells recirculated from the blood to the lymphatic system from where they were collected into the large thoracic duct before being returned to the blood. Thoracic duct lymphocytes (TDL) collected from one mouse and injected into the blood stream of another mouse would reappear in the thoracic duct of the recipient by 1 day after injection.

We will now describe another experiment. CBA mice were primed with horse red blood cells (HRC) and sheep red blood cells (SRC).

TDL from the CBA mice, adoptively transferred into syngeneic recipients and subsequently recovered from the thoracic duct one day after transfer, contained T helper cells for B cell responses to both horse red cells (HRC) and sheep red cells (SRC) as antigens. If, on transfer of the TDL, irradiated recipient mice were immunised with SRC, the TDL, recovered from these mice one day later, only provided help for HRC-specific B cells ie the SRC-specific helper T cells were not present in the TDL. This suggested that these T cells had been removed from the recirculatory pool by localised antigen.

If the experiment was repeated using B6 mice as the recipients (an allogeneic transfer), the TDL recovered one day after transfer and immunisation with SRC, contained T helper cells for both SRC and HRC responses ie the SRC specific T cells had not been removed from the recirculatory pool.

∏ Before you read on, see if you can explain this result?

These results were interpreted as indicating that antigen specific selection of donor T cells to SRC was dependent on the recognition of the MHC II determinants on lymphoid APC (antigen presenting cells) as self along with recognition of antigen. In other words, SRC specific T cells needed to 'see' both antigen and syngeneic MHC II on antigen presenting cells in the recipient for these cells to be retained in the mouse and subsequently provide help to SRC specific B cells in the animal. When transferred to the B6 mouse, the CBA T cells did not recognise the allogeneic environment of the recipient as self in spite of SRC being present and they remained within the recirculatory pool and reappeared in the TDL.

SAQ 4.6

Which of the following is/are correct or incorrect.

1) The congenic mouse strain B10.BR, haplotype H-2^k is constructed from mouse background strain C57BR (H-2^k) and an MHC donor B10 (H-2^b).

2) Sheep red cell specific thoracic duct cells, transferred from a (CBA x C57BL/10) F1 mouse into a recipient CBA/B10 congenic strain (see Table 4.1) mouse previously injected with the antigen, would not reappear in the thoracic duct of the recipient one day after transfer.

4.4.5 Adaptive differentiation of T cells and self-antiself recognition

Up to this point we have examined experiments in which adult T cells have been used and we have concluded that these cells collaborate (provide help) with B cells and are primed by antigen presenting cells only if there is some degree of syngeneicity in the MHC II molecules expressed on the surfaces of the interacting cells. The question that can be asked is, is there a self-self recognition or a self-anti-self recognition? Self-self recognition implies some sort of interaction between syngeneic MHC on the APC and the T cells together with a T cell receptor for antigen. Self-anti-self implies the presence of a receptor on the T cell for the MHC II molecules on the B cells or antigen presenting cells and a receptor for antigen. The latter conclusion would imply that the T cell did not have to express the same MHC as the APC.

self-anti-self recognition

A number of experiments using bone marrow chimaeras provided support for the self-antiself concept. Using this model some rather surprising results were obtained which on the face of it disagreed with the basic concepts of MHC restriction we have just discussed. Previous work had implicated the thymus as the maturation organ for T lymphocytes and bone marrow chimaeras were developed to study this concept.

bone marrow cells

Figure 4.9 illustrates a typical experiment. Cells were flushed out of the long bones of an (A x B) F1 mouse. These bone marrow cells contain multipotential stem cells that are the precursors for all haematopoietic cells including B and T cells. These cells were injected into an irradiated A-strain parent and the maturation of these stem T cells under the influence of the host thymus was studied. After about 8 weeks the mouse would be totally reconstituted by the donor cells and available for experiments.

After this time lymph node cells were recovered from the chimaera, the B cells removed and the remaining cells injected into an irradiated (A x B) F1 mouse along with B cells from a B strain mouse. The mouse was injected with SRC and the serum examined 7 days later for the presence of anti-SRC antibodies.

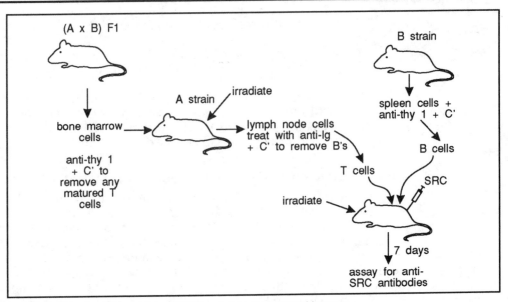

Figure 4.9 Adaptive differentiation of T cells using bone marrow chimaeras.

From the evidence discussed previously we would think that (A x B) F1 adult T cells would provide help to both A and B strain B cells. However, in this experiment there were no antibodies to SRC.

What had happened during the maturation process in the host thymus? Well, because the host environment was A, the cells in the thymus that promoted T cell differentiation only expressed A MHC antigens. The developing T cells had to learn what is self in the thymus and since only A MHC was expressed in the thymus they perceived this to be self.

Notice that the T cells still expressed the parental phenotype ie they expressed both A and B MHC Class I (T cells do not normally express MHC Class II) as they were derived from F1 mice. However, this phenotype is irrelevant to the learned recognition of self - it is the environment in which the T cells develop that is accepted as self - in this case the strain A environment. In this experiment, then, these T cells would only provide help to A strain B cells even though they expressed both A and B MHC I antigens.

adaptive
differentiation

This learning of self in an adopted thymic environment was coined 'adaptive differentiation' and Katz suggested that the T cells developed so-called cell interaction molecules on their surfaces while in the thymus. These molecules could interact or bind the self MHC antigens on the B cells or macrophages together with antigen. In other words, these findings support the idea of self-antiself recognition. This was proved to be correct with the identification of the T cell receptor (see below).

SAQ 4.7

A population of mice expressing the b haplotype for H-2K and D antigens (MHC Class I) and the k haplotype for I-A, I-E (MHC Class II) were irradiated and then reconstituted with cells expressing the d haplotype for H-2K and D and the d haplotype for I-A, I-E regions. Which one of the following statements is true?

1) The bone marrow chimaeras would not respond to antigen since the recipient cells and donor cells are allogeneic for MHC Class II. Hence no cell cooperation.

2) The mice would not respond to any antigens.

3) The mice would respond to a wide range of antigens since the donor cells include stem cells for both T and B cells. This means that cell cooperation between the donor T and B cells could take place resulting in antibody.

4) The T cells would collaborate with B cells expressing the d haplotype for MHC Class II.

5) The T cells would not provide any help to B cells in the bone marrow chimaera.

4.4.6 Interleukin 2 and the T cell receptor

Interleukin 2

In 1976 a momentous discovery was reported which rapidly led to identification of the T cell receptor (TCR). It was found that when T cells were stimulated *in vitro* they released a growth factor, T cell growth factor, now named Interleukin 2, into the culture medium which promoted the continued growth of T cells.

This discovery paved the way for researchers to clone T cells from single antigen specific, MHC restricted T cells. The basic procedure involved is the repetitive stimulation of T cells with 'antigen-pulsed, MHC-matched' APC's in culture which results in the expansion of T cells with specificity for the antigen-MHC pair. Individual antigen-MHC responsive T cells are then selected by cloning procedures and these can be expanded in the presence of IL-2 (see Chapter 6 for further details).

The most important aspect of this discovery was the ability to produce huge numbers of antigen specific, MHC restricted T cells at will. These was not technically feasible using antigen-driven T cell proliferation alone. This discovery quickly led to development of antibodies to the T cell receptor TCR which facilitated purification and characterisation of the TCR and then cloning of the cDNAs encoding the TCR molecules.

detection of T cell activation

It also led to a simple assay to detect T cell activation. When monoclonal T cells are presented with antigen-pulsed APCs they secrete IL-2 which can be measured. This leads us to our next experiment.

4.4.7 A single T cell receptor simultaneously recognises both MHC and antigen

At the time of the discovery of MHC restriction of T cells it had also been concluded that T cells simultaneously recognised antigen on the surface of APCs as well. The question that needed to be answered was, did the T cell have two receptors, one for the MHC and one for antigen or was there one receptor that bound both?

This question was elegantly answered by experiments reported by Marrack and Kappler in 1981 supported by evidence from others. The basic experiment was as follows. Using IL-2, a murine T cell clone was developed that was specific for an epitope on ovalbumin (OVA) and MHC restriction by H-2^k. These cells were fused to a **T cell hybridoma** thymoma cell line to produce a T cell hybridoma. The only characteristics of this hybridoma you need to know is that it possessed the same antigen specificity and MHC restriction as the parent T cell. If the hybridoma was presented with OVA on H-2^k APC, IL-2 would be released indicating that the T cell had recognised the antigen in the context of the MHC.

This hybridoma cell line was fused to a second T cell clone which was specific for an epitope on KLH and H-2^f restricted. The second order hybrid now possessed specificity for both OVA and KLH and had receptors for H-2^k and H-2^f.

After the construction of this 'double' hybrid, the experiment sought to answer the following questions: were there either individual receptors for each of the antigens and each of the MHC molecules (a total of 4 types of receptor) or were there only two receptors, one for OVA-H-2^k and the other for KLH-H-2^f. This was determined by presenting the hybrid with 3 different APCs as shown in Figure 4.10 and examining IL-2 release.

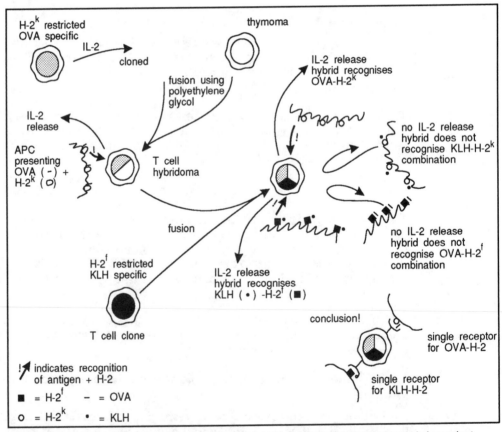

Figure 4.10 MHC restriction and antigen specificity are due to a single T cell receptor not independent receptors. The H-2 restriction depicted is due to the I-A region of the mouse H-2 (see text for details).

Π   Examine the results in Figure 4.10 and draw your conclusions before reading on.

The answer was unequivocal. The hybrid released IL-2 when presented with either OVA-H-2^k or KLH-H-2^f APCs but not when presented with OVA-H-2^f or KLH-H-2^k APCs. The most likely explanation for this result is that single T cell receptors bind both antigen and MHC.

SAQ 4.8

In an experiment on adaptive differentiation, A strain bone marrow cells were transferred into B strain irradiated recipients and after the chimaera has been established, spleen cells were tested for their ability to respond to hapten-carrier in cell culture (*in vitro*). There was no antibody response. Write down a brief explanation for this phenomenon. It might help you if you made a little drawing of the experiment and to think about the fate of the B and T cells in terms of what MHCs they will express on their surfaces and what MHC's they will recognise as self.

4.5 Antigen processing and presentation

4.5.1 Recognition of antigen by T and B cells

T cells see primary sequences

linear peptides

It has been well established that B cells are able to respond to native antigen whereas T cells respond to native or denatured antigen. This has led to the concept that B cells see tertiary sequences and T cells see primary sequences. Another important aspect of T and B recognition is that B cells can respond to small chemicals such as haptens, nucleic acids, polysaccharides, lipids and proteins. In direct contrast, T cells only recognise proteins and then only when they have been degraded to linear peptides. In other words, T cells only see linear amino acid sequences.

peptide binding clefts

By this time you will have realised that foreign antigens bind to MHC Class II molecules; the T cell receptor sees the complex; and is somehow activated by it. The ability to generate these complexes is the function of the antigen presenting cells (APCs). You will already know from your studies of the MHC, that protein antigens are first degraded into fragments or peptides and that these bind to peptide binding clefts in MHC molecules. The cleft has been identified on one MHC Class I molecule HLA-A2 and measures about 2.5nm long by 1nm wide and 1nm deep, sufficient to accommodate a peptide 8 - 20 amino acids in length. The binding cleft in MHC Class II molecules has not yet been described but is likely to be very similar.

The main properties of APCs then, are the ability to degrade these proteins, attach them to MHC II molecules and then present the complex on the surface for interaction with the T helper cell. Let us look at this mechanism in more detail.

4.5.2 Presentation of exogenous and endogenous antigen

exogenous antigens

endosomal pathway

Antigens derived from bacteria and soluble proteins such as KLH and OVA are called exogenous antigens because they are derived from outside the APC. These are taken into the cell by endocytosis, processed into peptides via an endosomal pathway, attached to MHC II Class molecules and expressed on the cell surface. The complex produced is recognised by T helper cells which bear on their surface a molecule called CD4. In fact, it is proper to say that all CD4$^+$ T cells are MHC Class II restricted. As we shall find out later, CD4 itself, also binds to the MHC Class II molecule.

endogenous
antigens

cytoplasmic
route

Antigens derived from inside the cell such as viruses or tumour antigens are said to be endogenous antigens. In the case of a virally infected cell, the peptides are produced during synthesis of viruses inside the cell. These endogenous antigens are processed into peptides via a cytoplasmic route, attached to MHC Class I molecules and expressed on the surface. The complex in this case is recognised by T cells, such as cytotoxic T cells (CTL), expressing CD8 molecules. CD8 itself binds to MHC Class I molecules.

A simple experiment emphasises these differences. Ovalbumin, when mixed in solution with macrophages, is taken up by the cells by endocytosis and peptides are found subsequently on the cell surface attached to MHC Class II molecules. However, if the gene for ovalbumin is transfected into the cell, the peptides are now found attached to MHC Class I molecules. In the first instance, the cell can activate CD4 T cells to produce IL-2; in the second instance, the cell is destroyed by MHC-I restricted CD8 CTL (see Figure 4.11).

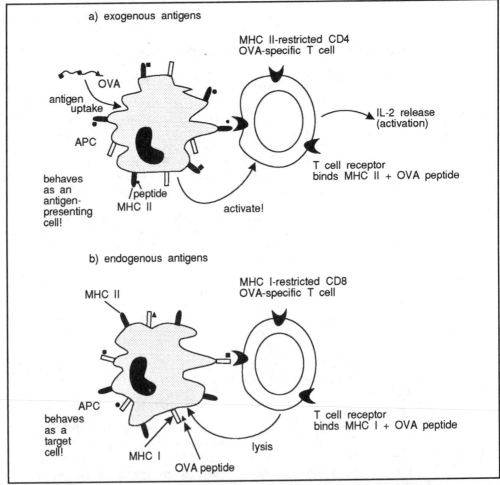

Figure 4.11 Presentation of exogenous and endogenous antigens. Note that with exogenous antigens, a) the antigen OVA administered exogenously is processed by the antigen presenting cell and portions of the antigen associate with MHC II. The MHC II peptides are then recognised by MHC II restricted OVA-specific. T cells and IL-2 is released. Note that the T cells has CD4 surface components. In b), we can follow the response to an endogenous antigen (OVA produced by genes within the cell). The endogenous antigen OVA associates with MHC I on the surface of the cell. This is recognised by MHC I-restricted OVA-specific T cell which responds by causing lysis in the target cell. Note that the T cell has CD8 surface components.

SAQ 4.9

Which one of the following statements is incorrect?

1) Exogenous antigens are those from which peptides are derived endosomally.

2) Endogenous antigens generally stimulate MHC I-restricted T cells.

3) Exogenous antigens such as OVA, after processing, can stimulate at least two sets of MHC II restricted T cells in an F1 mouse.

4) IL-2 would be released if A strain OVA-specific T cells, matured in an irradiated B strain mouse, are exposed to OVA presented by B cells from the strain A mouse.

5) Antigen processing precedes presentation.

Although we are not going to discuss cytotoxic T cells here since they are not directly involved in the control of antibody production, the subject of this chapter, it was necessary to define these differences. It is also important to remember that although it would seem that cells present antigen whether they express MHC Class I or II, we keep the term antigen presenting cells for MHC Class II expressing cells only. MHC Class I presenting cells are called target cells because they are lysed by CTL. We should also mention that a small percentage of CD4$^+$ MHC Class II restricted cells have been found to function as CTL as well as helpers; the reason for this is not yet clear.

4.5.3 Antigen processing requires time and metabolic activity

If APCs are pulsed with ovalbumin and then fixed immediately with glutaraldehyde, which renders the cells metabolically inert, these cells will not stimulate OVA specific T cells. However, if there is a 3 hour gap between uptake of OVA and fixing of the cells, then the APCs are capable of activating the T cells (Figure 4.12).

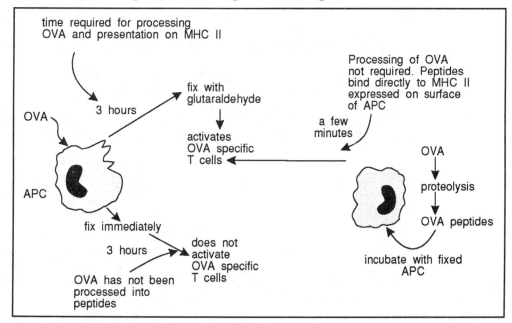

Figure 4.12 Antigen processing requires time and metabolic activity.

However, fixed APCs can effectively present antigen if OVA is first subjected to proteolysis (Figure 4.12). The 3 hour interval, appears to be required for the cell to process the protein and express it on the surface with MHC Class II molecules.

4.5.4 APCs present self peptides in competition with foreign peptides

It has recently been shown that the majority of MHC molecules expressed on the surfaces of APCs are occupied by peptide. In fact, MHC molecules are very unstable if they are 'empty'. We now assume that most MHC molecules expressed on the surface are, in fact, in association with a self peptide derived from inside or outside the cell.

Experiments have demonstrated competition between self (autologous) peptides and foreign peptides both *in vitro* and *in vivo*.

equilibrium dialysis

Equilibrium dialysis has been used to demonstrate competitive binding *in vitro*. The basis of the experiment is to place solubilised MHC Class II molecules inside a dialysis sac and place this in solutions of varying concentrations of fluorescent tagged peptides. With no MHC molecules in the dialysis sac, the peptide will finally attain the same concentration inside and outside the sac. However, MHC molecules inside the sac will bind peptide and this will increase the amount of peptide molecules inside the sac.

competitive binding

Addition of unlabelled autologous (self) peptides to the solution was found to reduce the binding of the fluorescent peptide indicating competitive binding by the autologous peptide.

competition by autologous peptides

Experiments *in vivo* have also demonstrated competition by autologous peptides. For instance, it has been shown that peptide 46-61 (amino acid sequence numbered from the N-terminal end) of hen egg white lysozyme (HEL) bound to I-A^k but not I-A^d, I-E^d or I-E^k. This HEL peptide only induced a response in I-A^k mice. A peptide from normal mouse lysozyme (a self protein in all mice!) ML46-62 shows a high degree of homology to the HEL peptide differing in 3 amino acid positions and it exhibits similar binding characteristics. As one would expect, however, it failed to induce a response in I-A^k mice because it is a self protein and, as we shall find out later, T cells do not normally respond to self peptide-MHC II complexes. However, if this peptide was injected into the experimental mice at the same time as the HEL peptide it prevented a response to the latter suggesting that it competes for the same sites on the MHC molecules.

Self peptides and foreign peptides have recently been eluted from affinity purified MHC II molecules and they appear to be heterogenous with an average molecular weight of 2-3kD. This development should lead to sequence determination of the peptides generated actually within the APCs and give us information on the structural features of self and foreign peptides that promote binding to MHC II.

You might well ask why self peptide-MHC complexes do not activate T cells *in vivo*. As we shall find out later, it is likely that there are no T cells capable of responding to such complexes, there are only T cells capable of responding to foreign peptide-MHC complexes.

Π Before you read on, try to work out the following problem. There are many potential peptides that could be generated from all the antigens in the environment but so few MHC II molecules. How can all these peptides be presented by so few MHC II molecules.

We will provide an answer in the next section.

4.5.5 Multiple peptides compete for each MHC binding site

MHC
molecules bind
many peptides

We are sure it will be obvious to you that the number of binding sites on MHC molecules encoded by a given haplotype is far less than those on antibody molecules. Mouse strains can express only a few versions of I-A and I-E molecules; indeed some mouse strains do not make I-E molecules. This suggests that each MHC molecule must bind many different peptides, even unrelated peptides. For instance, peptides derived from influenza virus haemagglutinin and hen egg white lysozyme have been shown to inhibit both MHC binding and T cell activation by an ovalbumin peptide. It has also been shown that two tetanus toxin peptides were recognised by all human donors irrespective of their MHC alleles suggesting a considerable degeneracy in MHC binding. It follows that even peptides with very little structural similarity must possess some common features that promote binding to the MHC.

degeneracy in
MHC binding

Bersofsky has suggested that amphipathic sequences are important, forming a helix with hydrophobic residues interacting with MHC and hydrophilic residues with T cell receptors (TCR). Other workers have demonstrated common amino acid residues in otherwise heterologous sequences. However the picture finally appears; we know the affinity of the binding is much less than that of antibody binding to antigen which is what you would expect but that TCR interaction with peptide-MHC stabilises the complex.

SAQ 4.10

In one investigation into the binding of peptides, experiments with single residue mutations of the HEL peptide demonstrated that residues 53 (tyrosine) and 56 (leucine) were contact residues for TCR whereas residue 55 (isoleucine) was a contact residue for the MHC II molecule. Which of the following pieces of evidence would lead to such conclusions?

1) Substitution of alanine at positions 55 or 56 resulted in no stimulation of HEL-specific T cells.

2) Mutants in which alanine was substituted in positions 53 and 56 bound to the MHC.

3) Changes in residue 57 or 58 had no effect on stimulation of the T cells.

4) An alanine (53) mutant failed to stimulate T cells.

5) An alanine (55) mutant did not compete with the native HEL peptide for MHC binding.

4.5.6 MHC binding to peptide dictates responsiveness

It should be apparent to you that unless proteolysis of a particular antigen results in a peptide which binds to at least one MHC Class II molecule, then the antigen will not be presented, T cells will not be activated and no response will ensue. The inability of antigenic peptides to bind MHC resulting in no response has been called the Determinant Selection theory. However, even if the peptide does bind to the MHC II, a response may not result if there are no T cells with receptors for that particular MHC-peptide combination. This suggests a limited T cell repertoire and is the basis for the Hole in the Repertoire Theory.

Determinant
Selection

Hole in the
Repertoire

4.5.7 The mechanism of antigen processing and presentation

Details of the actual processing and presentation of peptides in APCs are very incomplete at the moment. However, current ideas on this process are depicted in Figure 4.13 and describe the possible pathways of recycled and freshly synthesised MHC II molecules.

Use this figure to follow the description. The processing of endogenous antigens is shown on the left hand side of the figure, whilst we have shown the processing of exogenous antigens on the right hand side.

acidic environment

early endosome late endosome

Antigen taken up by APCs by either receptor mediated endocytosis, pinocytosis or phagocytosis are localised in a vesicle called an endosome. You can imagine that since this vesicle is formed by invagination of the cell membrane it will also contain MHC II molecules and a variety of other surface molecules and self polypeptides from the membrane-immediate environs of the cell. Proteolytic cleavage of the vesicle contents begins in an acidic environment using the proteases Cathepsin B and D. The endosome at this stage is termed an early endosome. At a later stage (late endosome) lysosomal enzymes also participate in an increasingly acid environment and a final lysosomal stage results in full degradation of the endosomal contents.

MHC II molecules are freshly synthesised in the endoplasmic reticulum (ER) along with MHC I molecules. The latter are though to pick up self peptides or endogenous peptides (such as viral peptides) while proceeding to the Golgi apparatus before being transported to the cell surface in a Golgi vesicle.

invariant chain

MHC II molecules appear to be treated differently. A new polypeptide called the invariant chain designated Ii has been proposed to bind to the newly synthesised MHC II molecules in the ER and remains associated during their passage through the Golgi. The binding of Ii is though to prevent binding of cytoplasmically derived self peptides to MHC II. Post-Golgi vesicles now fuse with either early or late endosomes and the Ii is proteolytically cleaved from the MHC II molecules thus exposing the binding clefts for association with the generated peptides. Some of the MHC II will bind exogenously derived foreign peptide while others will bind self peptides generated within the same endosomes. The endosomal vesicles are now presumably transported to the surface and expressed for interaction with CD4 T cells.

So you can see that, because of the presence of Ii, MHC II preferentially bind exogenously derived peptide while MHC I becomes saturated with endogenously derived peptides immediately after synthesis. Whereas the movement of MHC I from ER to cell surface takes a matter of minutes, that of MHC II takes up to 4 hours, most of this being taken up by residence in endosomal vesicles for processing.

directing recycling of MHC molecules

It is presently controversial whether MHC II molecules, recycled from the cell surface, are available to bind newly generated peptides and this may depend on the type of APC involved. One can imagine that, during the endosomal processing referred to above, peptide exchange would occur on recycled MHC II molecules. Ii molecules have recently been observed on the surface of B lymphoma cells which suggests that some invariant chains escape proteolysis. It is possible that these Ii are important in directing recycling of MHC II molecules through the endosomal compartments. Indeed, Ii molecules are being implicated in directing the transport of MHC II molecules from the Golgi to endosomes through an Ii receptor-mediated phenomenon.

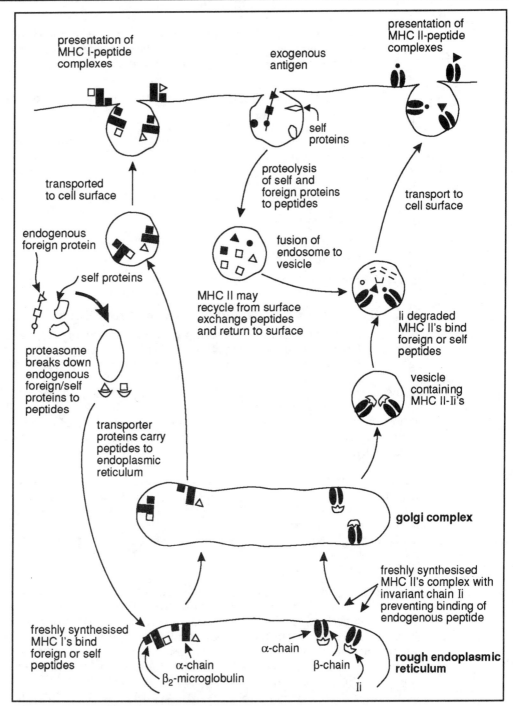

Figure 4.13 Antigen processing mechanisms. On the left hand side of the figure, we have shown the processing of endogenous antigens involving MHC I. On the right hand side, we show the processing of exogenous antigen involving MHC II. Ii = invariant chain.

⊓   It would be helpful for you to re-read this section and to draw your own flow diagram describing antigen processing mechanisms. Just draw a simple series of boxes describing each stage. Then compare it with our Figure 4.13.

SAQ 4.11

Complete the following table. Indicate by + or - the presence or absence of each feature.

Feature	MHC I	MHC II
Binds self peptides		
Binds exogenous antigens		
Involves invariant chain		
Proteases are involved		
ER to Golgi pathway		
Late endosome are involved		
Peptide binding cleft		

4.5.8 Cytoplasmic processing of antigens for MHC I presentation

proteasome

peptide transporters

It is convenient at this point to mention a very recent discovery regarding the supply of peptides for binding to newly synthesised MHC I molecules. It is now thought that genes in the Class II region encode glycoproteins which form at least part of a complex found in the cytoplasm called a proteasome and also other glycoproteins called peptide transporters. Antigens and self proteins are bound to the proteasome which degrades them in the cytoplasm. The peptide fragments are then carried by the peptide transporters across the membrane of the endoplasmic reticulum where they bind to newly synthesised MHC I molecules which are then moved to the surface of the cell (Figure 4.13).

Since both proteasome and transporter genes are found within the MHC, it has been suggested that the MHC is a complex of genes which control the processing and presentation of peptides to T lymphocytes. The cytokine gamma interferon has been found to simultaneously enhance the transcription of MHC, proteasome and transporter genes thus ensuring efficient presentation of peptides on Class I molecules.

4.5.9 Antigen presenting cells

The most important phenotypic characteristic of APCs is their expression of MHC II molecules. Some cell types always express MHC II - the constitutive expressors. These include Langerhans cells in the epidermis, interdigitating dendritic cells (IDC) found in T cell areas of lymphoid tissue and thymic medulla, B cells and cells of the cortical epithelium. Perhaps rather surprisingly, we often find MHC II expression on endothelial cells of capillaries and lymphatics and epithelial cells in such organs as the small intestine, the breast and trachea. Unless there is an as yet unknown role for MHC II, we have to assume that these cells may present antigen quite frequently!

gamma interferon

Notice the non-inclusion of macrophages in this class. It is generally accepted that resting resident macrophages express little if any MHC II but this can be induced by various bacterial products or by the cytokine gamma interferon (γ-IFN) produced by T lymphocytes. Once macrophages are activated, they express MHC II and are potent APCs. Apart from blood monocytes and lymphoid macrophages, these include marginal zone macrophages in the spleen, Kupffer cells lining the sinusoids of the liver and microglia in the brain. Macrophages are also capable of presenting antigen to B cells. We shall deal with this later.

Many other cell types can be induced to express MHC II *in vitro* by a variety of agents, the most important of which is IFN-γ. It is probably true to say that most cells can be induced until the stimulus is removed but there is very little *in vivo* evidence for this.

ΙΙ Before you read on why do you think all these cell types need to be able to express MHC II?

4.5.10 The role of APCs in clonal expansion of cytotoxic T cells

It is known that cytotoxic CD8 T cells require help from CD4 T cells for clonal expansion to proceed. Let us imagine a scenario where a virus has penetrated an epithelial cell and viral progeny are being produced. As we have said earlier, some of the viral proteins will be subjected to proteolysis in the cytoplasm, become attached to MHC I and be transported to the surface for recognition by MHC I-restricted CD8 CTL.

To generate memory CTL, you need clonal expansion of the virus-specific CTL and for this you need IL-2, the cytokine we referred to earlier. The general concensus is that this is not manufactured by CTL but by CD4 T cells. So, although a virus-specific CTL may recognise a viral peptide bound to MHC I on the infected epithelial cell it will not be activated and proliferate unless it receives the stimulus of IL-2. Since IL-2 acts across very short distances, the lymphokine has to be produced locally which implies that a CD4 cell has to be next to the CTL on the surface of the infected cell. However, in order to induce the CD4 cell to produce IL-2, the TCRs of the cell need to recognise and bind complexes of viral epitope + MHC II. You should now see why it is important for any cell to be able to express MHC II if infected by viruses.

It would appear that expression of less than 300 MHC-peptide complexes and ligation of about the same number of TCRs are sufficient to stimulate T cells. This number of MHC II molecules involved in the binding of a single peptide represents less than 0.1% of the total MHC molecules on an APC suggesting that a single APC is capable of presenting many different peptides simultaneously.

SAQ 4.12

Bone marrow (BM) cells matured in irradiated recipients which were then infected with virus were subsequently examined for their ability to lyse infected target cells (indicated by +). What is the most likely reason for the apparent failure of adaptive differentiation in these experiments?

BM cells	Irradiated recipient	Haplotype of target		
		A	B	C
A	(A x B) F1	+	-	-
B	(A x B) F1	-	+	-
(A x B) F1	(A x C) F1	+	-	-
(A x C) F1	(A x B) F1	+	-	-

1) Failure of strain A T cells to recognise B as self.

2) Lack of antigen presenting cells.

3) There were no carrier specific T cells in the A strain mouse.

4) You cannot develop fully allogeneic chimaeras.

5) The cells would have produced a response in the animal but not *in vitro*.

Summary and objectives

You have now completed a fairly difficult conceptual chapter where we have examined the evidence for cell-cell interactions in the production of antibody. We established the idea that many of the interactions between these cells are based on a pivotal recognition of self through the MHC and we examined this system in some detail. We discovered that each T cell can only recognise a single allelic product of either Class I (cytotoxic T cells) or Class II (helper T cells). We also showed that T cells have to learn to recognise self MHC and the important experimental evidence supporting the idea that there is a single T cell receptor for both MHC and antigen. We then developed the idea that T cells 'see' peptides which are a result of proteolysis of antigen within specialised cells called antigen presenting cells and that these cells also process and present self peptides in competition with the foreign peptides. We described the mechanisms by which peptide processing occurs in APCs and target cells which result in activation of helper and cytotoxic T cells respectively.

Now you have completed of this chapter you should be able to:

* describe reconstitution experiments which demonstrated T cell, B cell and antigen presenting cell requirements for the production of antibody;

* construct detailed diagrams of the genomic arrangements of the human and murine MHC and list the major structural features of Class I and II MHC molecules;

* define locus, allele, polymorphism, homozygous, heterozygous, syngeneic, allogeneic, linkage disequilibrium, congenic strain, graft versus host reaction;

* predict cellular interactions in Man and mouse given information on haplotypes;

* explain in detail what is meant by adaptive differentiation in bone marrow chimaeras and MHC restriction;

* describe the experimental approach that led to the concept of a single T cell receptor for MHC + antigen;

* describe in detail exogenous and endogenous antigen processing and presentation mechanisms;

* list the major criteria governing binding of self and foreign peptides to MHC molecules;

* list the types of antigen presenting cells and their characteristics.

Biology of T and B cells

Biology of T and B cells

Previously we have encountered B and T cells and, in Chapter 4 we explored the evidence for cell-cell interaction.

In this chapter we will extend our examination of cell-cell interactions in the immune response by describing the biology of B and T cells. We will begin by examining the isolation and characteristion of T cell receptors and by describing the genes which code for these receptors. We will subsequently move on to examine a large number of important cell surface molecules found on lymphoid cells. We will particularly emphasise their occurence, detection and function. We will then explore how appropriate T cells are selected and matured in the thymus. We will also explain how B and T cells are activated. Finally we will briefly examine the regulation of the immune response by suppressor T cells.

There is a considerable amount of information in this chapter. Much of it refers to cell surface components. Some of these have simple names, but many are given series of numbers. Especially important are CDs (cluster determinants or clusters of differentiation). It may be difficult to remember all of these but we have provided a glossary at the end of the text you may find useful. There is also a table at the end of the chapter which summarises much of the relevant information. We also suggest in the text that you make your own summary sheet of the various CD's, their occurence and function.

5.1 The T cell receptor

Determination of the structure and molecular genetics of the TCR has been the big discovery of the 1980's. It had to await the development of techniques that allowed the propagation of monoclonal T cells *in vitro* which itself was dependent to a great extent on the discovery of IL-2. In a monoclonal T cell population each T cell expresses the same T cell receptors on its surface and these will differ slightly from all other T cell clones. You will remember what B cell idiotypes were - unique sequences or determinants in the antigen binding site, often within the CDRs of the H or L chain. There are similar unique determinants in the TCRs of each T cell clone.

monoclonal T cells

5.1.1 Isolation and characterisation of the TCR using monoclonal anti-idiotype antibodies

cloned T cells

Once the T cells had been cloned, anti-idiotypic antibodies were prepared by immunising mice or rats with the cloned T cells and monoclonal antibodies were prepared from the spleens of these animals (details of this technique are given in the BIOTOL text ' Technological Applications of Immunochemicals'). Selected monoclonal antibodies were found which bound to the cells used to immunise the animals but not to any other T cells, indicating that they were specific for the idiotype on the TCR.

We will now look at the general procedure for isolating cell surface molecules from whole cells using the isolation of TCRs as an example. Use Figure 5.1 to help you follow the description given. Briefly, the T cells are surface labelled either with radioactive iodine or endogenously labelled with radioactive methionine, and then lysed with detergent. By this method, all the surface molecules including the T cell receptor molecules are separated from each other, are in solution, and most are radiolabelled.

The antibodies (Ab) are then added and the TCR-Ab complexes precipitated out of solution using Protein A or G attached to Sepharose beads. The latter two proteins are derived from bacteria and bind various subclasses of immunoglobulins. When attached to Sepharose beads, the proteins bind the antibodies via the heavy chains and so remove the complex out of solution.

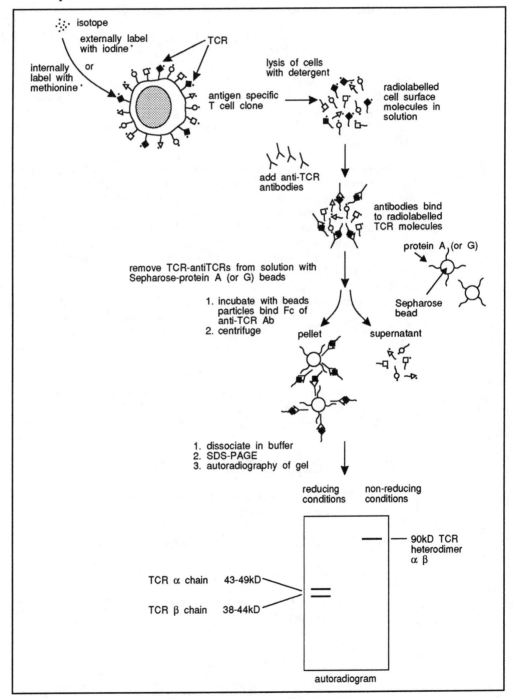

Figure 5.1 Isolation of TCRs from T cell membranes (see text for details).

∏   What other method could you have used to precipitate the complex? (see Figure
     2.5).

The method we hope you thought of is to use antibodies which bind the heavy chains
of the anti TCR antibodies in place of Sepharose - protein A.

The immune precipitate is dissolved in a suitable buffer and subjected to SDS (sodium
dodecyl sulphate)-PAGE (Polyacrylamide gel electrophoresis) to separate the
components by their molecular weight. The electrophoresis is carried out normally in
the presence of a reducing agent such as 2-mercaptoethanol which cleaves interchain
disulphide bonds (reducing conditions) or in the absence of reducing agents
(non-reducing conditions). The gels are then dried and exposed to X-ray film to produce
an autoradiogram of the separated polypeptide chains comprising, in this case, the T
cell receptor.

There are other ways to identify cell surface molecules such as receptors. An alternative
method, immunoblotting or Western blotting, is shown in Figure 5.2. Briefly, the cells
are lysed in detergent and subjected to SDS-PAGE. The separated proteins are then
transferred to a nitrocellulose membrane. This is done by laying a nitrocellulose
membrane onto the gel and allowing the proteins to be transferred to the membrane
rather like using a piece of blotting paper to suck up ink. This step is called blotting. The
nitrocellulose membrane with the blotted protein is then incubated with a non-related
protein to fill all the unoccupied sites on the membrane. This is called a blocking agent
as it blocks any further non-specific binding sites on the membrane. Then the
membrane is incubated with an IgG monoclonal anti-TCR antibody. These will bind
with the TCRs on the membrane. The antibodies bound to the TCR proteins are detected
using an anti-IgG antibody-enzyme complex. The TCR proteins will show up as
coloured bands when the nitrocellulose membrane is incubated with the enzyme
substrate.

SAQ 5.1	An experiment was done by both of the methods described in Figure 5.1 and 5.2 using the same monoclonal anti-TCR antibodies. No bands were detected in the Western blotting experiment. Which of the following is the most likely reason for this?

1) The TCR proteins were not transferred to the nitrocellulose membrane.

2) The anti-TCR antibodies could not bind to the TCR bands on the
   nitrocellulose membrane.

3) The enzyme system was not sensitive enough to detect the TCR bands.

4) The TCR proteins were destroyed during SDS-PAGE.

5) The second antibody did not bind to the TCR-anti-TCR antibody complex on
   the membrane.

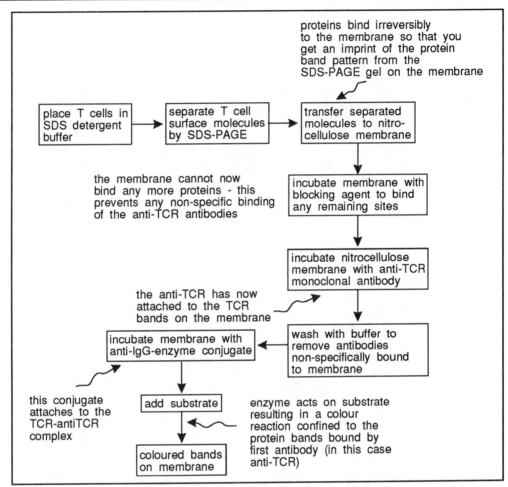

Figure 5.2 Western blotting. Note that the boxes and straight arrows indicate the experimental steps, the wavy lines indicate a brief commentary of what is happening at each stage (see text for further details).

5.1.2 Basic structure of the TCR

The results obtained from this and related investigations showed that human TCRs from a majority of T cells are heterodimers of 2 polypeptide chains α and β, of molecular weight 43-49kD and 38-44kD respectively, joined together by a disulphide bridge. A drawing of a TCR is given in Figure 5.3. The α chain is an acidic glycoprotein possessing some N-linked oligosaccharides, the β chain a basic or neutral glycoprotein with both N-linked oligosaccharides and sidechains rich in high mannose-glycan. Both chains can be divided into variable and constant regions. These receptor molecules are always found physically associated with a complex of at least 5 polypeptides, the CD3 complex (see below).

The variable regions (domains) are at the N-terminal ends of the chains and are furthest from the cell membrane. They are about 100 - 120 amino acids in length and include a disulphide bonded loop. They have structural similarities to Ig variable domains and each possess three regions of hypervariability (CDRs = complementarity determining regions). The α and β variable domains interact with each other and include the binding

α and β chains

variable regions

disulphide bonded loop

sites for recognition of MHC + peptide. It is thought that both chains contribute to binding both MHC and peptide.

constant
regions

homology with
Ig constant
domains

The constant regions of both α and β chains are from about 140 to 180 amino acids in length and include an interchain disulphide link. Each constant region consists of a disulphide-looped constant domain which exhibits homology with Ig constant domains, a small hinge region, a hydrophobic transmembrane portion of about 20-25 amino acids and a short cytoplasmic tail. The transmembrane sequence includes a lysine residue in the β chain and a lysine and an arginine residue in the α chain which may be important in interactions with acidic residues at similar sites in components of the CD3 complex with which the TCR is closely associated. The structure of the T cell receptor is shown in Figure 5.3. Examine the figure carefully.

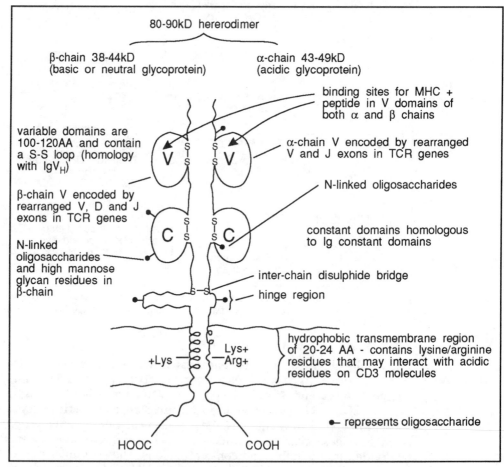

Figure 5.3 Basic features of the T cell receptor (TCR2). AA = amino acids (see text for a further description).

5.1.3 There are two major forms of the T cell receptor

TCR-2

The majority of T cells express the αβ TCR (see 5.1.2). This is generally referred to as the TCR-2 receptor and is found on almost all CD4 or CD8 expressing T cells.

TCR-1

A minority of T cells, generally CD4⁻, CD8⁻, however, do not express TCR-2. These αβ-negative T cells express TCR-1 which consists of a γ and a δ chain. TCR-1 receptors are always associated with CD3 in a similar manner to TCR-2. The two chains of TCR-1 are structurally very similar to the α and β chains with the same pertinent features. A comparison of the amino acid sequences show close homology between the β and γ chains and the α and δ chains.

The biological functions of this subset of T cells has not yet been determined. However they appear to be concentrated in tissues of the gastro-intestinal tract especially intra-epithelial sites. Most TCR-1 expressing cells do not express CD4 or CD8 and do not appear to be MHC-restricted. Some of them have been found to have non-specific CTL activity against tumour cell lines and others antigen specific CTL activity. It has been suggested that they may be active against common microbial antigens that invade through the epithelium.

intra-epithelial
sites

5.1.4 Interactions between the TCR and MHC + peptide

Each V domain possesses 3 CDRs similar to immunoglobulin heavy and light chain V domains. CDR 1 and 2 are encoded by the V exons and CDR 3 is encoded by V and J in the α chain gene and V, D and J in the β chain gene, very similar to Ig L and H chains respectively (see 5.2.2). In a similar fashion to immunoglobulins, a total of 6 CDRs then can contribute to the binding of TCR to MHC + peptide.

The manner in which the TCR chains interact with MHC + peptide is not known. However, we do know that CDR 3 contributes much more variability to the receptors than CDR 1 or 2 for the same reasons as for immunoglobulin genes. Since the MHC haplotypes in any one individual are not subject to change and most of the variability presented to the TCR will be based on the peptide, it is probable that CDR3 of each TCR chain will interact with the peptide and the other CDRs with the MHC determinants.

SAQ 5.2

Complete the table showing presence by a +, absence by a -.

Structural feature	α-chain	β-chain
Binds peptide		
Acidic glycoprotein		
Homology to Ig		
CD8$^+$ cells		
Transmembrane lysine + arginine		
CD4$^+$ cells		
V, D and J exons		
Binds MHC		

| **SAQ 5.3** | Complete the table indicating presence by +, absence by -. |

Feature	TCR-1	TCR-2
Antigen specific		
CD8$^+$		
Heterodimer		
MHC restricted		
CD4$^+$		
CD3$^+$		
Restricted specificity		
Most common receptor		

5.2 T cell receptor genes

5.2.1 Identification of TCR genes

subtractive
hybridisation

The identification of TCR genes (Figure 5.4) was achieved by the use of a method called subtractive hybridisation. This was based on the correct assumption that the majority of mRNA transcripts in T cells are identical to those in B cells. In fact these cells differ in less than 1% of their mRNAs. The mRNAs from membrane polysomes of an antigen specific T cell hybridoma were used to prepare cDNAs using reverse transcriptase. These were then hybridised with B cell mRNAs and the DNA-RNA duplexes separated from the nonhybridised cDNAs. The remainder of the identification procedure is similar to that described for Ig gene rearrangement in Chapter 3 (Figure 3.1). In other words, we were left with cDNAs which were copies of mRNAs which were unique to T cells. Copies of these cDNAs were then synthesised using radioactive nucleotides so that radioactive cDNAs were produced. DNA was then extracted from B cells, liver cells and T cells, treated with restriction enzymes to cut them into specific fragments and the fragments separated by gel electrophoresis. These separated fragments were then melted (strands separated) on the gel and challenged with the radioactive cDNAs. This allowed identification of the fragments of DNA which carry the nucleotide sequences of the unique T cell messenger RNAs. In effect, the radioactive cDNAs were used as probes to identify genes uniquely expressed in T cells. This technique to identify specific genes is usually referred to as Southern blotting.

Using this technique TCR genes could be identified in DNA fragments prepared from liver (L), B cells (B) and T cells (T). This procedure showed that in T cells, the TCR genes were on different fragments than those in liver and B cells. In other words, TCR genes were rearranged in T cells.

Once the TCR genes were identified in this way, DNA fragments bearing the TCR genes were isolated, cloned and that nucleotide sequences determined.

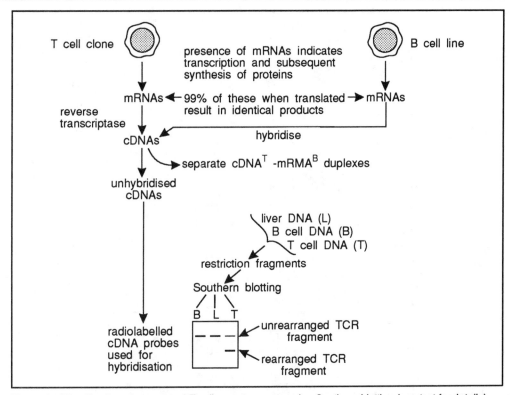

Figure 5.4 Identification of rearranged T cell receptor genes using Southern blotting (see text for details).

5.2.2 Genomic organisation of TCR genes

T cell receptor genes

The T cell receptor genes have been cloned and show similarities to immunoglobulin genes. Only the murine β locus has been completely sequenced. The genomic organisation of Man and mouse TCR genes are very similar and is shown in Figure 5.5. One rather unusual feature is the location of the δ chain locus within the α chain locus.

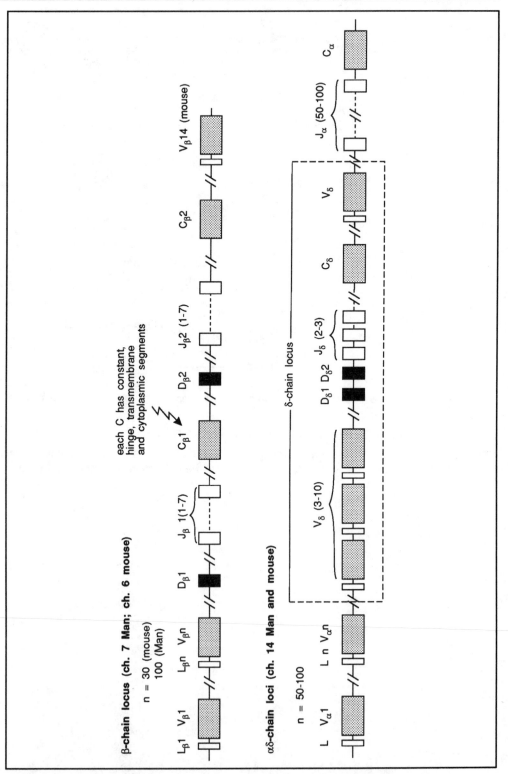

Figure 5.5 Genomic organisation of the murine and human α, β and δ TCR genes (not to scale), ch = chromosome.

∏ How would you expect rearrangement of the α chain gene to affect that of the δ chain gene (examine Figure 5.5)?

It should be self-evident that in the rearrangement of α-chain gene fragments to produce functional α-chain gene, the δ chain gene is eliminated.

The rearrangement mechanisms are similar to those of Ig genes including the use of 12 and 23-base pair spacers and nonamers and heptamers. In fact, you can transfect TCR genes into pre-B cell lines and rearrangement occurs.

The potential diversity of TCR genes has been calculated to be equivalent to that of the Ig genes. Although the numbers of V segments is lower than for Ig genes, there is a much greater number of J segments. In addition junctional diversity is thought to be more widespread and involves more flexibility than for Ig genes and N region additions occur in both α and β genes.

no somatic hypermutation in TCR genes

Although there is a small number of V segments in γ and δ chain genes, once again a variety of unusual rearrangements involving junctional diversity can result in enormous potential diversity. You should note, however, there is no somatic hypermutation in TCR genes. It is interesting to note that α chain gene rearrangements will eliminate the δ chain gene and suggests that δ chain gene rearrangement must occur earlier in development or the TCR1 and TCR2 expressing cells are separate lineages.

You will realise, of course, that both CD4 and CD8 T cells express TCRs. It is interesting to note that the same gene pool is used by both these populations of cells even though they are restricted by different MHC class molecules.

SAQ 5.4

Which of the following is/are correct?

1) The murine β locus has been completely sequenced and was found to have more than 50 J segments.

2) TCR α and β genes are similar to Ig H and L chain genes in that both junctional diversity and N region additions can occur to create additional diversity.

3) Rearrangement of the α-chain VDJ exons would eliminate the δ-chain locus.

4) CDR3 of both α and β chains contributes more than CDR1 and 2 to the diversity of TCR2 and probably binds the peptide of the MHC-peptide complex.

5.3 Cell surface molecules on lymphoid cells

5.3.1 CD classification

Lymphoid cells possess many different molecules on their surfaces and many of these have been identified using many monoclonal antibodies (cluster of antibodies) from laboratories across the world and have been designated CD meaning cluster determinant or cluster of differentiation followed by a number. Typical examples which

cluster determinant

you have already come across are CD4 and CD8 which identify helper and cytotoxic T cells respectively. The number of cell surface specificities now numbers about 80. When a new structure is tentatively identified by antibodies, it is initially assigned a workshop number (eg CDw49) prior to a proper CD classification. You will come across quite a few of these CD antigens which are involved in cell adhesion, may act to transduce signals leading to cell activation and which are differentially expressed during certain stages of lymphocyte development. You will need to remember quite a few of these! It would be a good idea to construct a summary sheet as you read through the following sections dealing with CD antigens.

We suggest you use the following headings.

CD Group   **Occurence**   **Structure**   **Function**

Remember that the major cell markers and the activities of the lymphoid cells is given in a summary table at the end of the text. There is also a glossary of abbreviations. Use these to help you.

5.4 The CD3 complex

5.4.1 Structure of CD3

CD3 complex

The CD3 complex is found on T helper and T cytotoxic cells. As we mentioned earlier, TCRs are always found associated with the CD3 complex which is thought to transmit activation signals to the interior of the cell once TCR has formed a terniary complex with MHC-peptide. CD3 consists of at least five different integral proteins closely associated with each other and with TCR. Expression of TCR and CD3 is co-dependent ie one is not expressed without the other.

Figure 5.6 shows the principal features of the CD3 proteins. There are five distinct proteins in the complex. These include two glycosylated members, (the 25-28kD γ chain and the 20kD δ chain) and three nonglycosylated members, the 20kD epsilon (ε) chain, the 16kD zeta (ζ) chain and the 21kD eta (η) chain. The first three chains exist as monomers in the complex. In contrast, the zeta chain is expressed as a homodimer in a majority of T cells and as a heterodimer with the η chain in others. Recent evidence also suggests that the η chain may exist as a homodimer on some T cells. The various TCR

TCR isoforms

isoforms known to exist are depicted in Figure 5.6. Examine this figure carefully, it contains a lot of information, but it provides a useful summary.

TRAP

There is yet another protein involved with the TCR-CD3 complex, the T cell receptor-associated protein (TRAP), also designated by a ω. This 28kD molecule is thought to be involved in the assembly of the TCR-CD3 complex but does not itself appear on the T cell surface.

transducing signals

The actual arrangement of the CD3 chains within the TCR-CD3 complex is not defined at present. However, in contrast to the short cytoplasmic tails of the TCR chains, all of the CD3 proteins possess long enough cytoplasmic tails to be capable of transducing signals to the cytoplasm of the cell. It is therefore, presently accepted that once TCR has bound the MHC-peptide, the CD3 complex transduces the activating signals to the cytoplasm of the cell.

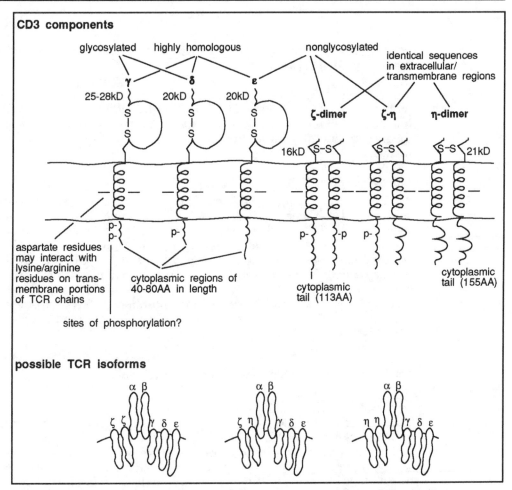

Figure 5.6 The CD3 complex and TCR isoforms. AA = amino acid. αβ refers to TCR subunits.

5.4.2 Activation of the T cell

The activation process within the cytoplasm and nucleus of the T cell is extremely complex and is still being elucidated. We shall only mention the principal steps. Following binding of MHC-peptide by the TCR, the CD3 complex transduces a signal by activation of phospholipase C that catalyses the hydrolysis of a membrane associated phosphatidylinositol bisphosphate to inositol triphosphate (IP_3) and diacylglycerol (DAG). The released IP_3 enters the cytoplasm and promotes the release of calcium from an intracellular store. Together with other signals we will discuss later, these products are thought to activate cellular proto-oncogenes in the nucleus which then leads to transcription of the Interleukin 2 gene and the IL-2 receptor gene. This leads to the release of IL-2 from the cell and expression of the IL-2 receptor - essential steps in the activation and proliferation of T cells. We will come back to this in Chapter 6.

phospholipase C

Interleukin 2 gene

IL-2 receptor

5.5 Other accessory molecules involved in cell interactions and T cell activation

promotion of
adhesion

The binding of TCR-CD3 to MHC-peptide is not the complete story as there are other molecules on the surface of T cells and APCs that either promote adhesion between the interacting cells or possess co-stimulatory activities. We shall briefly mention some of them. You should be aware that this is a new area of research activity and there are many gaps in our knowledge.

5.5.1 CD4 and CD8

CD4 molecules

nonpolymorphic
regions of the
MHC II

All T helper cells express CD4 molecules, 55kD transmembrane proteins, on their surfaces. In other words, CD4 is a marker for T helper cells. The activation of T helper cells requires the presence of CD4. These molecules bind to the nonpolymorphic regions of the MHC II and possibly enhance the binding of TCR-CD3 to the MHC-peptide especially when the affinity of the TCR is low. It may also be physically associated with TCR-CD3 and be involved in signalling during T cell activation. The lymphocyte specific activation protein tyrosine kinase $p55^{lck}$ is associated with CD4 and CD8. Tyrosine kinase $p55^{lck}$ catalyses the phosphorylation of tyrosine residues within target proteins. In Man, CD4 is also found on macrophages and it is the receptor for the AIDS virus.

CD8

CD8 is found on all CTL and suppressor T cells and has a similar function to CD4. CD8 binds to the nonpolymorphic regions of MHC I molecules.

5.5.2 CD2

CD2,
spontaneous
rosettes,
leucocyte
function-
associated
antigen-3,
anti-CD2
antibodies

CD2, also called T11, LFA-2, Tp50, Leu 5, is a 50kD glycoprotein present on T cells and on natural killer cells. It was originally identified as the receptor for sheep red cells which were found to form spontaneous rosettes with T cells and afforded a means for separating these cells. It has two principal functions. Firstly, it is an adhesion molecule which binds leucocyte function-associated antigen-3 (LFA-3), a 55-70kD glycoprotein expressed on the surfaces of many cell types. Secondly, it is a signal transducing molecule and some anti-CD2 antibodies can promote proliferation of T cells and lymphokine secretion. It may be associated with TCR-CD3 and be involved in TCR-mediated activation.

∏   Can you think of a way to use sheep red cells to separate T cells?

Since the T cells will be surrounded by a rosette composed of bound red blood cells, this in effect increases the density of the T cells. If the rosetted cells are placed on a suitable gradient and centrifuged they will separate from non-rosetted cells. You will see an example of this technique resulting in the separation of T cells from B cells in Section 8.2.1.

SAQ 5.5

Complete the table indicating presence of specific molecule on T cell subsets with a +, absence with a -.

Surface molecule	T helper cell	T cytotoxic cell
TCR		
CD3		
CD4		
CD8		
CD2		

5.5.3 CD28

B cell activation marker

CD28 is also known as Tp44 (90kD molecular weight). CD28 is expressed on all CD4 cells and about 50% of CD8 T cells. It has recently been found to bind a B cell activation marker B7 (BB1) on B cells (remember, these are APCs) and is involved in regulating cytokine production.

5.5.4 CD45 leucocyte common antigen

This is a major cell surface glycoprotein found on T cells (also called T200) and B cells (also called B220) and other white cells. The gene contains 34 exons which, with alternate splicing, results in up to eight different proteins of molecular weight 180-220kD. There is a lot of speculation on the possible role of this molecule in activation although there is little doubt that it is involved. Cell mutants lacking CD45 fail to proliferate after antigenic or other stimulation through CD3. Various isoforms identified by monoclonal antibodies are found on cells in different stages of activation. You will meet two of these, CD45R and CD45RO, later.

5.5.5 Other activation molecules

extracellular matrix

You will notice that all the accessory molecules (except CD45) referred to so far have adhesion properties (they bind ligands on other cells) and co-stimulatory functions. We should mention just a few others with similar properties. The T helper cell expresses CD26 and VLA (Very late antigen)-3 that both bind collagen in the extracellular matrix (tissues outside the cell), VLA-4 and 5 that bind fibronectin and VLA-6 that binds laminin. Notice, these do not interact with the APC but with the extracellular matrix. These interactions probably also promote directed movement of these cells through the tissues.

5.5.6 Cell adhesion molecules

LFA-1

Intercellular Adhesion Molecule-1

VCAM 1

Additional molecules on T cells possess primarily an adhesion function but are also involved in controlling the migration of the T cells. The principal molecule is LFA-1, composed of two chains CD11a and CD18. It belongs to a family of molecules called Integrins and is expressed on all bone marrow derived cells. It appears to be very important in cell adhesion as anti-LFA-1 antibodies block T cell activation and CTL killing. Its principal ligand is the Intercellular Adhesion Molecule-1, (ICAM-1) also designated CD54 but it also binds ICAM-2 found on many cell types. T cells also express ICAM-1 that interacts with LFA-1 on the APC and other interacting cells. Finally, VLA-4 on T cells binds the ligand Vascular Cell Adhesion Molecule, (VCAM-1) on APCs and other cells. We shall come to some of these in Chapter 7. We have summarised these interactions in Figure 5.7. Examine this figure carefully because it provides a lot of information.

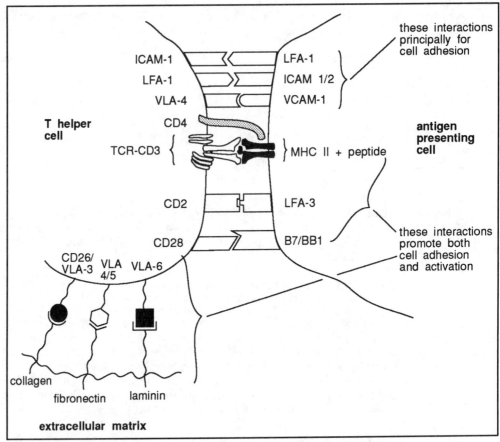

Figure 5.7 Molecules involved in cell interactions between T helper cells and antigen presenting cells.

⫪ Notice the interactions between cytotoxic T cells and target cells are very similar. Can you remember the main differences?

Did you remember them? The major differences are that CTL express CD8 instead of CD4 that binds to the monomorphic region of MHC I molecules rather than Class II molecules. As far as we know, all the other interactions exist between cytotoxic cells and target cells. However, only 50% of CTL express CD28 in contrast to all CD4 cells.

So you see, the interaction of T helper cells and APCs and, indeed other cells, is a fairly complex affair and it will be a long time before the whole story is unravelled. It should emphasise to you the importance of cell-cell contact in activation processes.

| SAQ 5.6 | Match the items in the left hand column with those in the right hand column using each item only once. |

1) CD11aCD18 (LFA-1)   a) Sheep red cell receptor

2) LFA-3   b) B7 (BB1)

3) Collagen   c) MHC II

4) CD28   d) Fibronectin and VCAM-1

5) VLA-4   e) CD8

6) CD4   f) ICAM-1

7) MHC I   g) CD26

5.6 Maturation of T cells in the thymus

Earlier we examined the evidence for adaptive differentiation in bone marrow chimaeras. It was concluded that immature T cells or pre-T cells were able to develop, in the thymus, receptors to MHC molecules expressed in the thymic environment irrespective of the parentage of the T cells. In other words, they learned a new self! Since we already know the structure of the TCR and we also know that TCR genes rearrange in the thymus to produce MHC-peptide specificities, let us just briefly examine the major steps occurring in the thymus resulting in the release of mature T cells into the peripheral pool. Again we have to warn you that this is an extremely complicated and controversial aspect of immunology. Therefore it is sufficient at this stage for you to get a general picture regarding the role of the thymus. Use Figure 5.8 to help you follow the sequence of events.

5.6.1 The thymic cortex - positive selection

pre-T cells

thymic cortex

apoptosis

Pre-T cells develop from multipotential stem cells in the bone marrow as do B cells and other lymphoid cells. Early in their development, they migrate to the thymic cortex where they undergo rearrangement of the TCR genes and also intense proliferation. The cells expressing TCR-1 first appear followed by TCR-2 expressing cells. At this stage the cells have also developed CD1, 2, 3, 5 and 7 on their surfaces, CD1 only being expressed during development. TCR-2 expressing cells then express both CD4 and CD8 (called double positives) and are exposed to either MHC I or II on thymic epithelial cells. Those cells that are able to bind molecules of either MHC class with low to high affinity are 'rescued from cell death' by unknown mechanisms. If a CD4$^+$ CD8$^+$ cell binds to MHC I via its TCR, it is presumably destined to be MHC I restricted and will now stop expressing CD4. Similarly those cells binding to MHC II become MHC II restricted and lose CD8. The remaining cells, probably 95% of the total cells, die by a process known as apoptosis which involves fragmentation of DNA. This whole process is known as positive selection - all the living cells at this stage recognise the MHC displayed in the thymus. The TCR-2 cells then differentiate to either CD4$^+$ or CD8$^+$ cells.

Π Can you now think back to bone marrow chimaeras and think how allogeneic stem cells come to be able to adopt the host as self?

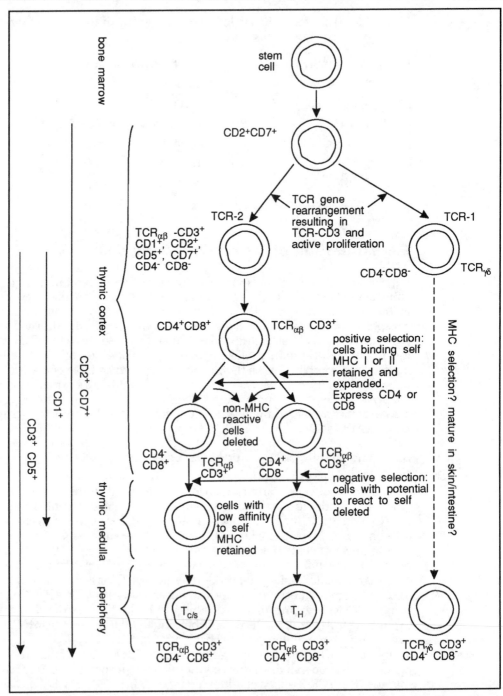

Figure 5.8 T cell maturation in the thymus. The arrows down the left hand side of the figure indicates the duration of the surface markers CD1, CD2, CD3, CD5 and CD7.

positive selection This positive selection is, of course, what happens to stem cells in bone marrow chimaeras. In the experiments we discussed, even though expressing allogeneic MHC I on their surface, some of them underwent TCR gene rearrangement that resulted in a TCR which recognised the host MHC and were positively selected. In other words, among those 95% of cells which died, TCR rearrangements would possibly have resulted in their positive selection in an allogeneic host!

5.6.2 The thymic medulla - negative selection

thymic medulla

The positively selected cells now move to the thymic medulla where they come into contact with a different set of antigen presenting cells - principally bone marrow derived macrophages and dendritic cells. The potentially autoreactive cells, those binding with medium to high affinity to either the MHC I (the CD8 cells) or MHC II (the CD4 cells) undergo apoptosis and are destroyed. Only the cells with low affinity TCR receptors for the self MHC molecules are retained and exit the thymus into the periphery as immunocompetent $CD4^+$ or $CD8^+$ cells ie helper T cells or cytotoxic T cells respectively. This process is called negative selection.

negative selection

The development of TCR1 T cells has not been not elucidated - we are not sure whether they are selected by similar mechanisms involving the MHC. These cells may mature extra-thymically, maybe in the skin and intestine.

∏ Below we have drawn summary scheme of the changes in the surface components as cytotoxic T cells (T_c) develop from stem cells. Check that this is consistent with the information given in Figure 5.8.

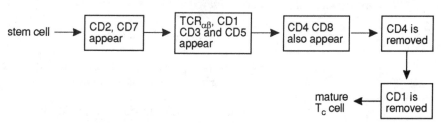

The scheme is essentially correct.

∏ Now draw yourself a scheme showing the changes to the surface markers in helper (T_H) T cells.

Your scheme should be essentially the same except that CD8, not CD4 is removed (see fourth box in the above scheme).

∏ Note that the scheme illustrated in Figure 5.8 has two important stages in the selection of appropriate T_c, T_s and T_H cells. Make a summary of what these are and what they do.

We hope you have identified these stages. The first stage of selection takes place in the thymic cortex and selects only those T cells which will bind self MHC I or II. The second stage of selection takes place in the thymic medulla and leads to the removal of T cells which bind rather strongly (high affinity) with self MHC I or II. Thus we are left with a population of T cells which can recognise self MHC I or II but do not bind very strongly with these self components. The next question is how do we select T cells which will respond to self peptides? We will deal with this in the next section.

5.6.3 Self and foreign peptides

You might ask where do self peptides come into this story? Again, we have to say we do not really know the answer. However, we have already concluded that MHC molecules are expressed stably on the surface of cells only when complexed to a peptide, in many cases, self peptides. It is thought that the medullary APCs present the whole spectrum of self peptides that are normally generated by these cells. For instance, B cells during normal activities take in self molecules from the immediate micro-environment and generate endogenous peptides from cellular metabolism which are expressed with MHC I. Similarly, the macrophages and dendritic cells generate sets of peptides which are expressed with both classes of MHC.

If the developing T cells in the thymic medulla bind to these MHC-self peptide complexes with anything but low affinity they are destroyed ie any T cells that could be activated by such a complex are deleted. T cells that bind with low affinity to such complexes are not activated and hence are not a potential threat to the host. So, once the T cells leave the thymus they will not be activated by self MHC-self peptide complexes.

low affinity T cells can respond to foreign peptides

So how are these low affinity T cells able to respond to foreign peptides and initiate responses? The idea is that if the self peptide is replaced by a foreign peptide of slightly different structure, some of the T cells, although exhibiting low affinity binding to MHC-self peptide, will bind the foreign peptide-MHC complex with medium to high affinity and thus will be activated.

SAQ 5.7

Which of the following are true and which are false?

1) Positive selection is a process by which immature T cells are selected by binding to host MHC on bone marrow derived cells.

2) TCR gene rearrangement precedes CD4, CD8 differentiation in TCR-1 and TCR-2 expressing cells.

3) The development of T cells in the thymus is antigen independent.

4) Self peptides expressed on B cells may differ from macrophage derived self peptides.

5) The only major differences in the surface characteristics of TCR-1 and TCR-2 expressing cells are the T cell receptor chains.

5.6.4 The use of transgenic animals

transgenic mice

Further understanding of what is going on in the thymus has been facilitated by the use of transgenic mice. These are animals which have had a foreign gene introduced into their germline, such genes being called transgenes. The technique for introducing transgenes is briefly as follows.

Female mice are induced to produce many eggs by hormonal injection and are mated immediately. A few hours later, the fertilised eggs are recovered from the oviducts before cell division has begun. The eggs are then microinjected with the male pronucleus along with about 200 copies of linearised DNA of the transgene. The success rate of getting genes incorporated by this technique is about 25%. If the scientists succeed in injecting the DNA before cell division occurs and the DNA is successfully incorporated into the genome, then copies of this DNA will be found in the cell and become part of the germline ie the transgene will be passed on to the progeny. (Details of this technique are given in the BIOTOL texts 'Biotechnological Innovations in Animal Productivity' and 'Strategies for Engineering Organisms').

Remember that TCR genes are rearranged in mature T cells (see Figure 5.4 and associated text) so that they express functional α and β chains. Once this gene rearrangement has taken place, it suppresses further rearrangement. In other words, other α and β chain alleles are suppressed, a process called allelic exclusion. Thus any one particular T cell will produce its own TCRs all composed of the same α and β chains. With this information attempt the next intext activity.

Π If the transgenes were rearranged TCR α and β chain genes derived from a single T cell clone, then most if not all the T cells in the transgenic mice would express the same TCR. Can you explain why before you read on?

In positive selection experiments using this system, it could be shown that TCR needs to bind to the MHC in the thymus for the developing T cells to survive. For instance, transgenes encoding a TCR which was H-2^d restricted, when place in H-2^d mice resulted in mature T cells detected in the peripheral lymphoid tissues. If placed in H-2^b mice no mature T cells developed. This is because virtually all developing T cells in transgenic mice express this one transgenic TCR due to allelic exclusion (the rearranged gene blocks other rearrangements) and they will only be positively selected in mice expressing the MHC allele for which the TCR is restricted.

The system has also been useful in studying negative selection. In one experiment, transgenic mice were constructed using a transgene encoding a TCR with specificity for the H-Y molecule only found on cells of male mice. It was found that only female mice had normal numbers of T cells expressing this transgene in the thymic medulla and the peripheral lymphoid tissues. Apparently, the T cells in the male mice were exposed to self peptides derived from the H-Y molecule along with MHC on APCs in the thymic medulla and were deleted.

5.7 T cell and B cell markers

Although we have introduced you to quite a few cell surface molecules on T lymphocytes they cannot all be classified as T cell markers since they are found on other cell types as well. This is the case with most of the adhesion molecules we mentioned. So let us just list those which serve to identify T and B cells. Add these to your growing list of surface components.

5.7.1 T cells

As well as TCR and CD3 all T cells also express CD2, CD5 and CD7.

5.7.2 B cells

B cell markers Early in B cell development, the B cell expresses B220 and CD19. Both these markers persist on mature B cells although the B220 is lost after progression to the plasma cell. CD19, CD20, CD22 are the major markers present on all B cells along with membrane immunoglobulin and MHC Class I and II. We know little about the role of these markers in B cell activation. CD19 (95kDa) is thought to associate with membrane IgM and function to regulate B cell adhesion. It may be involved in B cell activation. CD20 (36kDa) may regulate the uptake of calcium by the B cell and is probably involved in activation. CD22 (135kDa) is probably an adhesion molecule. There may be expression of CD74, the membrane form of the invariant chain referred to earlier. The majority of B cells also express CD21 which is the receptor for a complement component C3d and for Epstein Barr Virus in Man, and CD35, the complement receptor for C3b. They may also express Fc receptors for IgG and IgE.

As for T cells, we have to admit that we know very little about the complexities of the involvement of B cells surface molecules and of their role in activation.

A minority of B cells express CD5, a marker found on all T cells. This subset of B cells has been referred to in Chapter 3 and is responsible for the production of polyreactive antibodies that react with self antigens. They are prominent in some autoimmune diseases.

IgM receptor complex

The antigen receptors on B cells are IgM and IgD. Recent evidence suggests that the monomeric IgM receptor is complexed with other molecules B29 (Ig-β) and MB-1 (IgM-α). This complex is probably the B cells equivalent of CD3. Memory B cells express other isotypes.

Remember that B cells express many other molecules including adhesion receptors and these are referred to elsewhere in the text.

5.7.3 The use of markers to separate T and B cell subpopulations

panning

Polystyrene surfaces avidly bind proteins including antibodies and this property is used in separation of lymphocytes on anti-Ig coated plastic surfaces, a process called panning. As an example, if you wish to separate B cells from T cells, you would incubate the cell suspension in a polystyrene petri dish coated with anti-Ig antibodies. These would bind the antibody receptors on B cells and the T cells could be washed off with buffer. The B cells can be recovered with more vigorous washing with normal serum which competes with the B cells for the bound antibodies. Often the anti-Ig antibodies will bind the Fc portion of the Ig. Such a preparation would separate a range of B cells (that is any with Fc on their surface). If we use an anti-Ig antibody which binds the antigen binding site of the surface antibody (that is, we use an anti idiotype antibody) then this will select only those B cells which are producing particular idiotype antibodies.

In some experiments an indirect method may be preferable. In this case, the cells are incubated with the particular antibody, say anti-CD4, and then the cells are incubated in a petri dish coated with anti-Ig (Fc) antibodies. The dish would retain the CD4$^+$ T cells.

A more recent development for separating T and B populations involves the use of magnetic beads coated with marker antibodies. The beads are mixed with the cells in a culture tube and the tube is exposed to a magnet. The beads carrying the isolated cells can thus be separated from the nonbinding cells.

Another approach is to use marker antibodies directed against cells you do not require and then lyse them with complement. You would need to make sure that the antibody you use actually fixed complement!

SAQ 5.8 Which one of the five protocols (1-5) depicted in Figure 5.9 would result in a reasonably pure population of cytotoxic T cells.

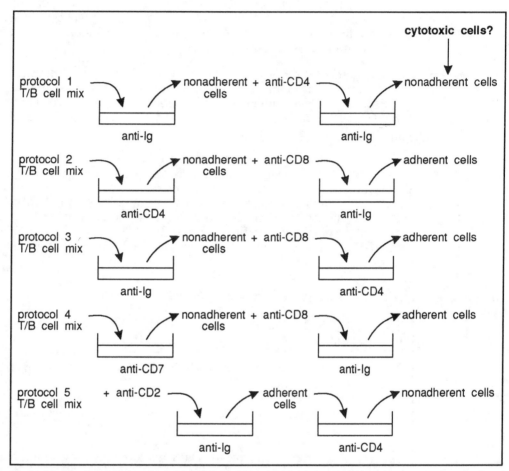

Figure 5.9 Panning technique (see SAQ 5.8). Note that anti-Ig binds cells with surface retained antibodies, anti-CD4 and anti-CD8 refer to antibodies which bind CD4 and CD8 respectively.

5.8 T helper and B cell activation

5.8.1 Assays for B cell activation

proliferative assay

B cell activation is detected either by a proliferative assay or by antibody production. B cell proliferation resulting from antigenic stimulation or activation by mitogens is measured by adding radiolabelled thymidine to the cells and measuring the amount incorporated into DNA. This is usually proportional to the rate of cell division. B cells can be stimulated by poke weed mitogen and by anti-immunoglobulin (antibodies which bind with B cell surface receptors).

Briefly, the cells to be assayed are recovered from the tissues or blood, counted, resuspended in tissue culture medium and equal numbers are placed into individual wells of a microculture plate. Serial dilutions of antigen, mitogen or anti-immunoglobulin, usually in microgram quantities, are then added and the cells are placed in an incubator kept at 37°C and flushed with 5-10% carbon dioxide. After 3-4 days, radiolabelled thymidine is added to each well and a few hours later, the cells are harvested from the wells onto fibre filters which are then placed in scintillation fluid for counting in a scintillation counter. Construction of a graph of counts per minute versus dose of antigen/mitogen will indicate the dose range under which the B cells respond and the optimal dose for the response.

measurement of antibody production

Current methods for measurement of antibody production in supernatants of activated B cells are radioimmunoassay (RIA) or the enzyme-linked immunosorbent assay (ELISA). Alternative methods include haemagglutination and complement mediated lysis. (These techniques are described in detail in the BIOTOL text 'Technological Applications of Immunochemicals').

haemolytic plaque assay

Measurement of single cell production of antibodies is by the haemolytic plaque assay. This detects the number of cells in a B cell population producing antibodies of a particular isotype. The antigen can be covalently coupled to red cells and these are mixed with lymphocytes in a semisolid medium such as agar. Antibodies secreted by the B cells diffuse from the cells and binds to the antigen on the red cell surface. Addition of complement induces lysis in these cells and the area of lysis (plaques) around each cell is counted giving an estimate of the numbers of B cells producing the antibodies.

SAQ 5.9

Which one of the following statements is correct?

1) Molecules (markers) present on all B cells include CD5, CD19, CD20 and CD22.

2) T cell molecules that promote adhesion and activation include ICAM-1, CD28, CD2 and CD4.

3) B cell proliferation can be measured by the enzyme-linked immunosorbent assay.

4) The hemolytic plaque assay is a quantitative assay for antibodies.

5) The majority of B cells are susceptible to infection with Epstein Barr Virus.

5.8.2 B cell activation by T-independent antigens

Although a majority of B responses to for example, protein antigens, depend on T helper cells, it is known that athymic mice can respond to a variety of antigens. These antigens have been called T-independent antigens (TI antigens) and they are subdivided into two groups depending on their structure and cytokine dependence.

T-independent antigens

TI-1 antigens are totally T-independent. An example is the lipopolysaccharide (LPS) or endotoxin found in Gram negative bacteria. At high concentrations this is a polyclonal activator promoting proliferation of B cells independent of the immunoglobulin receptors. At low concentrations LPS selectively activates LPS specific B cells through the Ig receptor. There are no known cytokines involved, in most cases, no isotype switching (we said this was probably cytokine dependent in Chapter 3), no somatic hypermutation and no memory are involved.

TI-2 antigens include such molecules as dextrans, pneumococcal polysaccharides and Ficoll. These antigens are polymeric and present repeat structures to B cells. These only stimulate through the conventional Ig receptor and crosslink the Ig receptors thus activating the cells. As for TI-1, the principal antibodies produced are IgM. It is thought that specialised macrophages may take up these antigens and subject them to partial hydrolysis and then expose them on the surface to B cells. It is thought that some cytokines may be involved.

5.8.3 The priming of T helper cells

Although we have implicated macrophages as being required for the activation of T cells and we know that B cells are potent APCs, there appears to be a very special type of cell that primes T cells. It is called the interdigitating dendritic cell (IDC) or the immunostimulatory cell and is found in the paracortex of lymph nodes which is an area where T cells tend to congregate. IDCs express high amounts of MHC I and II on their surfaces but lack receptors for both Ig (Fc receptors) or complement (C3b receptors) and are nonphagocytic. It seems probable that they are derived from Langerhans cells of the skin. The latter also express MHC I and II but in addition express receptors for both Ig and C3b. These cells could therefore pick up antigen in the skin and travel up the lymph stream to the nodes and present antigen to T cells in the paracortex. One imagines that on their travels in the lymph they lose receptors for Ig and C3b and express high levels of MHC II. This may maximise presentation of the peptides derived from the antigen they picked up in the skin.

interdigitating
dendritic cell

Langerhans
cells

Once these T cells are primed, they can interact with other APCs and become activated.

5.8.4 B cells as antigen presenting cells

B cells are very specialised antigen presenting cells since, for the most part, they endocytose antigen in an antigen-specific manner in contrast to other APCs which take up antigen nonspecifically. The antigen is initially bound by the surface immunoglobulin receptors and the whole complex is endocytosed into an endosomal vesicle. Processing of the antigen then proceeds and the B cell expresses a number of peptides along with MHC II on the surface. Memory T cells bind to the peptide-MHC II complexes through their TCRs and become activated (see below). The T cell simultaneously primes the B cell mainly through the action of cytokines. We will discuss these subsequent events later. Notice that each B cell can potentially receive help from a number of peptide specific T helper cells depending on how many different peptides it presents on the surface and how many MHC II alleles bind a single peptide (Figure 5.10).

Π From Figure 5.10 can you 1) identify any of the self peptides 2) determine the epitope specificity of the antibody receptors and 3) provide evidence for MHC molecules binding structurally unrelated molecules?

An examination of Figure 5.10 indicates that some of the self peptides are derived from antibody receptors on the B cell. For instance, MHC II^a (second complex from top of figure) is binding an antibody fragment and so is MHC II^c (sixth from top). The figure also shows that the antibody is specific for epitopes contained within peptide E. The MHC molecules are clearly binding many different peptides. For instance MHC^b is binding peptide C, peptide D and 2 different self peptides.

It is important to note that B cells are not what we might call antigen crunchers. They do not seem to be capable of major proteolytic activity and receive partly digested antigen from other cells. For instance, it is known that macrophages infected with viruses, as well as presenting viral peptides to T cells also 'present' degraded viral antigen to B cells in a non-MHC restricted manner. Another cell type, the 'follicular dendritic cell (FDC)', resident in the follicles of the cortex of lymph nodes, is devoid of MHC II but bears receptors for IgG and C3b which can trap antigen. Complexes of antigen and antibodies called iccosomes are presented to B cells in germinal centres by these cells and the B cell endocytoses the complex, processes it and presents the resulting peptides for stimulation of T helper memory cells. These in turn activate the B cells into a secondary response.

follicular dendritic cell

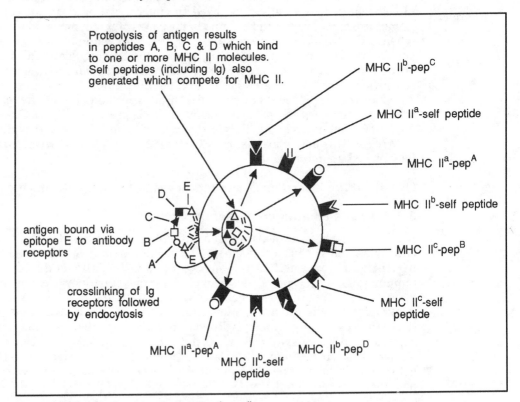

Figure 5.10 The B cell as an antigen presenting cell.

SAQ 5.10

Indicate by a + which cells express the phenotypes or functions listed in the left hand column.

Function/phenotype	B*	T*	La*	IDC*	FDC*
Classical APC					
FcR$^+$, CR$^+$, MHC I/II$^+$					
Antigen specific					
FcR$^+$, CR$^+$, MHC I$^+$II$^-$					
Iccosome production					
Activates unprimed T cells					
Activates primed T cells					
Activates primed B cells					

*B = B cells; T = T cells; La = Langerhans cells; IDC = interdigitating dendritic cells; FDC = follicular dendritic cells; FcR = Fc receptor; CR = complement receptor.

5.8.5 Maturation of the antibody response-selection of high affinity B cell clones

It is now thought that most B cells die within a few days of leaving the bone marrow. To survive they must meet up with antigen. Even then, a majority of activated B cells make no contribution to the immune response. The activation and selection of high affinity cells may take place in the following way.

class switching and somatic mutation

The B cells may be activated by T helper cells within the paracortex whilst part of a three cell complex, the third cell being the IDC. So the IDC activates the T cells and the T cell activates the B cell by producing cytokines. The B cell then migrates across to the germinal centre within the follicle. The B cell divides many times and the progeny undergoes class switching (producing a different isotype) and somatic mutation under the influence of the T cell products. Most of the B cells die and are taken up by macrophages as 'tingible bodies' unless they are rescued by being brought into contact with antigen presented by follicular dendritic cells (FDC) as described earlier.

anti-CD40 antibodies

Yet another B cell surface molecule, CD40, may be involved in this interaction with FDCs as it has recently been shown that anti-CD40 antibodies + anti-IgM can prevent apoptosis in the B cells.

With low antigen levels, the B cell must be expressing high affinity Ig receptors to be able to associate with the FDC ie high affinity clones must be produced, in many cases, as a result of somatic mutation such high affinity clones are now selected by antigen on FDC and these clones expand under the influence of T cell cytokines. The B cells will either migrate to the medulla or other lymphoid sites and secrete antibodies or will become high affinity memory cells.

We will complete the story of B and T cell activation when we consider the roles of cytokines in Chapter 6.

5.8.6 Resting and memory T and B cells

Progress is being made on identifying T and B cells that have entered the memory cell pool. Some distinguishing features are helping researchers to differentiate between resting/naive and memory T and B cells.

T cells

The naive T cells must be primed by IDC and on activation produce IL-2 and low amounts of IFN-γ. They express low levels of surface CD2 and a particular isoform of CD45, the leucocyte common antigen, CD45RA. They also express low levels of two homing receptors CD44 and L-selectin. These homing receptors are thought to bind to ligands on endothelium in small blood vessels thus directing the cells expressing the homing receptors to selectively extravasate from the blood stream to tissues or lymphoid organs. We shall discuss them in some detail in Chapter 7.

T cells are thought to enter lymph nodes directly from the blood through specialised blood vessels called post-capillary venules (PCV's). More about those in Chapter 7 also.

Memory T cells, on the other hand, can be activated by any APC and produce a wide variety of cytokines apart from IL-2 and IFN-γ. They express high levels of CD2 and a different isoform of CD45 called CD45RO. They may express high levels of both homing receptors and also of LFA-1 and LFA-3. Many of these cells appear to traffic through the tissues and not enter directly into lymph nodes from the blood stream.

B cells

Naive or resting B cells are generally low affinity cells that do not recirculate. They express IgM and IgD and low levels of complement receptors and the homing receptor L-selectin.

Memory B cells that have undergone affinity maturation and class switching are high affinity cells found in the recirculatory pool. They have lost surface IgM and IgD and express other isotypes. Complement receptor expression is high and so is that of the homing receptor L-selectin.

5.8.7 Comparison of T and B cells

Π Before you proceed any further, see what you know about the biology of T and B cells. Construct a comparative table showing the major characteristics of T and B cells under the headings 1. Development/Maturation site 2. Functional heterogeneity 3. Antigen receptor 4. Subclasses 5. Cell surface markers 6. Unprimed cell characteristics 7. Memory cell characteristics and 8. Major function. When you have completed it compare it with Table 5.1 in Section 5.11.

SAQ 5.11	Which one of the following is a characteristic for B memory cells?

1) They express idiotypes on their IgM receptors.

2) They are part of the recirculatory pool expressing low levels of L-selectin.

3) They have undergone somatic mutation and express high affinity IgM and other isotypes.

4) They have probably been exposed to antigen on FDCs in an MHC II unrestricted manner.

5) They are part of the CD5 lineage.

SAQ 5.12	Complete the table indicating possession of property with a +.

Property	Helper T cells	B cells
MHC II restricted		
Interacts with FDC		
Primed by all APCs		
Expresses MHC II		
Zeta/eta chains		
Positive selection		
Expresses MHC I		
Cytokine production		
Native antigen		
Haemagglutination assay		
Memory cells		
CD45+LFA 1+		
Homing receptors		

5.9 Regulation of the immune response by suppressor T cells

The deletion of clones of self reactive T cells in the thymus is not the only mechanism by which tolerance to self is maintained. Experiments in 1970 by Gershon and others suggested that non-responsiveness to self antigens may also be due to the presence of suppressor T cells which negatively regulate the activities of helper T cells and B cells. Thus such cells could suppress autoreactivity to self antigens not presented in the thymus.

As yet, there are many unanswered questions about suppressor T cells. In this and the following sections we will examine some recent discoveries.

5.9.1 Demonstration of suppressor T cells

Mice were injected with tolerogenic doses of antigen and spleen cells were transferred from these mice to syngeneic mice which were then given an immunogenic dose of the antigen. The antibody response was found to be depressed. However, if the spleen cells were treated with anti-T antiserum plus complement to destroy T cells prior to transfer, there was no depression of the response. This experiment suggested that at least some of the T cells in the spleen exposed to the tolerogenic doses of antigen were responsible for suppressing the antibody response in the recipient animal which had been exposed to an immunogenic (antibody stimulating) dose of the antigen. These T cells were called suppressor T cells (T_s).

5.9.2 Major characteristics of suppressor T cells

Suppressor T cells have been reported to be highly sensitive to X ray-irradiation and cyclophosphamide. They slowly disappear following thymectomy in adult animals; this procedure does not affect T helper cells. They express CD8 and a marker encoded by a gene which was thought to be situated close to the I-E region in the H-2 complex called the I-J region. The markers carried on the suppressor T cells were called I-J molecules. However, a gene for this product has never been identified.

Some suppressor T cells are able to bind antigen and it has been possible in some experiments to enrich spleen cells for suppressors using an antigen column. This contrasts sharply with T helper cells which can only see processed antigen bound to MHC. However, whether all T suppressors can bind antigen in this way is doubtful.

Reports, using mice, suggest that the activities of suppressor cells are antigen specific and MHC restricted and soluble factors released into the supernatants of these cells are heterodimers consisting of a 45kD antigen-binding chain disulphide linked to a 28kD chain bearing the I-J determinant.

It has been suggested from some experiments that these soluble factors could be antigen receptors released from the cell surface. Recent reports indicate some success in the cloning of suppressor T cells and these express TCR-CD3 complexes which have been reported to be released into the supernatant. However, there are discrepancies in the molecular weights of the putative factors and the TCR. The mechanism of recognition by suppressor cells and whether it involves antigen processing is still not known.

5.9.3 Mechanisms of immune suppression

suppressor circuits

The activation and regulation of suppressor T cells is complex and largely unsubstantiated. It possibly involves distinct subsets of T cells and suppressor factors. It is thought that during the initiation of a response an APC activates a CD4$^+$ suppressor inducer cell (similar to a T helper cell) which then activates a CD8$^+$ suppressor T cell. This cell releases an antigen specific suppressor factor which binds to a third cell, the acceptor suppressor T cell. This is antigen specific and releases antigen specific suppressor soluble factors but may also release antigen non-specific inhibitory factors which can suppress both helper T cells and B cells generally in a non-MHC restricted manner.

Π   Use the above description to make yourself a flow diagram of the sequence displaying how immune suppression works. Begin with the antigen on an APC and end with the suppression of B and T_H cells. Then check your diagram with ours.

The sort of information you should have included in your diagram is:

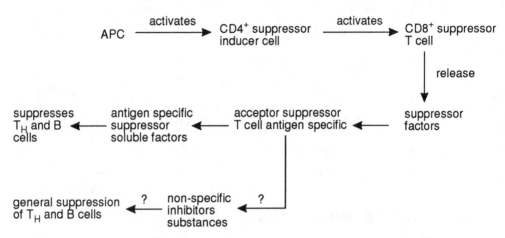

Obviously there are a lot of questions about the nature of the various factors involved in this sequence. There is however some evidence that suppressor mechanisms may involve anti-idiotype recognition.

A particular strain of mice produces a single idiotype of anti-azobenzenearsonate when it was immunised with azobenzenearsonate.

If mice of this strain were pre-treated with antibodies which react with this idiotype (that is they react with the antigen binding sites of anti-azobenzenearsonate antibodies) and then immunised with azobenzenearsonate, they then produce antibodies of quite different idiotypes. In other words the anti-idiotype antibodies suppressed the production of the antibodies it was capable of reacting with.

But how may such anti-idiotype antibodies be produced? Remember that B cells internalise antibody-antigen complexes and process, not only the antigens into peptides, but also the antibodies themselves. This results in peptides derived from antibody variable domains (including the idiotype) being presented on the MHC II. Thus this might activate idiotype specific T cells which in turn would provide help to these idiotype-expressing B cells. In this way, we would produce anti-idiotype antibodies. It could be that a similar mechanism could operate for the production of idiotype specific agents by T suppressor cells.

We will not pursue this story any further at this stage as it is presently confused and incomplete. We can however anticipate that we will eventually unravel this complex matter.

SAQ 5.13	Choose which of the following describe characteristics of both T helper and suppressor cells.

1) They bind antigen alone.

2) They may be MHC II restricted.

3) They may be idiotype specific.

4) They secrete regulatory molecules.

5) They are derived from $CD4^+CD8^+$ cells.

5.10 B cell tolerance

The main two mechanisms for tolerance to self antigens are by clonal deletion of T cells and by the production of suppressor T cells which prevent the response of T helper cells. The latter mechanism may be operative when self antigen is present at extremely low levels. For example, thyroglobulin is present in serum at extremely low levels. Individuals possess B cells with receptors for this molecule but the B cells are not activated due to suppression of the thyroglobulin specific T helper cells by antigen specific suppressor T cells. When antigen levels are high, however, such as is the case for serum albumin or the ABO blood groups, there are no responsive B cells. This could be due to clonal deletion of B cells during development when exposed to high levels of persisting antigen or to induced clonal anergy of B cells. In the latter case, the B cells are thought to have a block in the expression of antigen receptors and cannot respond to the self antigen.

5.11 A comparison of T and B cells

In this chapter we have provided you with a considerable amount of information about T and B cells. Here we provide you with a summary of this information.

Characteristics	T lymphocytes	B lymphocytes
Development/ maturation site	Early stages in the bone marrow but must pass through thymus to become immunocompetent.	Bursa of Fabricius in chickens; bone marrow probable site in other animals.
Heterogeneity of function	Possess functional heterogeneity being classed as helper T cells (Class II restricted and including cell mediated immune effector cells), cytotoxic T cells (Class I restricted) and suppressor cells.	B cells are functionally homogeneous - they all have a single function - to produce antibodies.
Antigen receptor	Complex receptor consists, on most cells, of a disulphide linked heterodimer of an α chain (43-49kD) and a β chain (38-44kD). This is the TCR-2 receptor. A minor T cell population express the $\gamma\delta$ TCR-1 receptor. The TCR always found associated with CD3. TCR does not bind native antigen only linear peptides resulting from proteolysis.	Receptor is a monomeric form of IgM and IgD. Memory cells express other isotypes. Receptor is found associated with additional chains in a complex.
Subclasses	There are TCR-1 and TCR-2 T cell subpopulations. The TCR-2 are further divided into T_H, T_C and T_S.	B cells have two major subpopulations CD5$^+$ and the majority of B cells CD5$^-$.
Cell surface markers	All mature T cells express MHC Class I, CD2, CD3, CD5 and CD7. TH also express CD4 and CD28; T_C and T_S express CD8 and 50% CD28.\n\nAlso express LFA-1 (on all bone marrow derived cells), CD45R (all leukocytes). Immature T cells express CD1.	All B cells express CD19, CD20 and CD22. Most mature Bs have CD21 (Complement receptor 2 for C3d) which binds Epstein Barr virus and CD35 (Complement receptor 1 for C3b). Most have Fc receptors for IgG (Fc$_\gamma$RII, CDw32) and IgE (CD23). All express MHC Class I and II, LFA-1, CD45R (B220). B cells also express CD24, CD40, CD72 and CD73 found on other cell types.
Unprimed cells	Express low levels of CD2, CD44 and L-selectin. Express CD45R and produce IL-2 and IFN-γ.\n\nActivated by interdigitating dendritic cells and enter lymph nodes directly from blood.	Generally express low affinity receptors of IgM/IgD isotypes. Express low levels of complement receptors and L-selectin.
Memory cells	Express high levels of CD2 and CD45RO, CD44 MEL-14, LFA-1 and LFA-3. Produce many different cytokines and can be activated by an APC. Cells traffic through tissues to nodes.	Cells have undergone affinity maturation and class switching to high affinity receptor bearing cells expressing isotypes other than IgM or IgD. Express high levels of complement receptors and L-selectin. These cells recirculate.
Function	Control cell mediated immune reactions and T/B responses.	Production of antibodies in humoral immunity.

Table 5.1 Comparison of T and B cells.

Summary and objectives

In Chapter 4 we looked at the evidence for cell-cell interactions in antibody production. In this chapter we examined the molecule basis for these cell interactions ie the receptors and ligands on the surfaces of B cells, antigen presenting cells and T helper cells that promote activation of T and B cells. You now have a considerable knowledge of the T cell receptor-CD3 complex and how other CD molecules help in cementing the adhesion between interacting cells. Many of these same molecules may be involved in the actual activation of T and B cells and we shall return briefly to this topic in Chapter 6. However, we know little about the role of these molecules.

We also extended our discussion from Chapter 4 and learned about the structure and role of the TCR in recognition of self and how the sequential mechanisms of positive and negative selection results in a repertoire of antigen reactive cells. We also briefly examined an alternative to thymic deletion, that of antigen specific suppression.

We further emphasised the importance of the cell phenotypes by looking at markers for T and B cells and how these can be used for separation of the cells. We also devoted a lot of attention to the priming and activation of T and B cells, again based on cell-cell recognition phenomena, and found that there are a variety of accessory cells crucial to the activation of T and B cells.

Now you have completed this chapter you should be able to:

- describe procedures for isolating and identifying cell surface molecules;
- list the major structural features of the TCR-2 T cell receptor;
- differentiate between TCR-1 and TCR-2 cells;
- describe the major features of the TCR genes;
- distinguish between helper and cytotoxic T cells by surface markers;
- identify cell surface molecules and their ligands involved in interactions between T helper cells and APCs;
- describe the main events in thymic education;
- choose the correct protocol for purification of T or B cells by panning;
- compare the structural and function characteristics of helper T cells and B cells;
- list methods by which B cell activation is measured;
- distinguish between the phenotypic and functional characteristics of accessory cells in T and B cell activation;
- differentiate between nonprimed and memory T and B cells;
- describe the major features of T cell mediated suppression.

Cytokines - the intercellular messengers

Cytokines - the intercellular messengers

6.1 Introduction

6.1.1 Active supernatants contain cytokines

cytokines are
found in cell
supernatants

In the 1960's, reports started to appear in the scientific literature on so-called active supernatants or conditioned medium from activated cells demonstrating the presence of soluble factors which mediated various effects on immune cells. The first demonstrations of such activity were reported in peritoneal exudate cells from immunised guinea pigs which inhibited the migration of macrophages in culture dishes when exposed to the specific antigen. The active component was termed macrophage

macrophage
migration
inhibition factor
(MIF)

migration inhibition factor (MIF). It was further shown that supernatants of these peritoneal exudate cells after exposure to antigen exhibited the same effect as the cells, thus suggesting that a soluble component had been secreted into the supernatant. In similar experiments, it was also demonstrated that the supernatants contained another soluble factor which induced cell division in T cells ie a mitogenic factor. Both of these

mitogenic factor

factors were antigen non-specific but were produced by antigen stimulation of T cells.

At about the same time another factor was reported in supernatants of mixtures of allogeneic cells ie a mixed lymphocyte culture (We will deal with this in Chapter 8) which substituted for T cells in a secondary response by primed B cells to the antigen

T cell replacing
factors

sheep red cells *in vitro*. A plethora of factors were now being reported in supernatants of activated cells and they were collectively called T cell replacing factors.

In this chapter we will begin by examining how cytokines are prepared before describing their general properties. We will then focus onto Interleukin 2, the most fully studied cytokine. Then we will examine the other cytokines with particular emphasis being placed on their synthesis, structure and activities in T and B cell activation.

6.1.2 Preparation of active supernatants

The normal procedure was to collect either murine spleen cells or human peripheral blood cells and place these in culture medium together with a suitable stimulus. This could be antigen, mitogens such as plant lectins, phytohemagglutinin, concanavalin A (con A), phorbol myristic acetate (PMA) or allogeneic cells. These were cultured under standard conditions for 48hr and then the supernatant was recovered from the cells by centrifugation. It was then filter sterilised and examined for its effects on various immune reactivities of T and B cells. These supernatants were found to contain antigen specific T helper and suppressor factors and antigen non-specific factors some of which promoted various activities of immune cells, others which inhibited such activities.

These factors were all difficult to characterise as they were present in extremely low amounts, possibly nanomolar quantities, in the supernatants and hence could only initially be detected by their biological activity.

6.1.3 Purification of cytokines from active supernatants

recovery of cytokines from supernatants

The supernatants were preferably prepared in serum free culture fluid to eliminate contamination problems by serum proteins in subsequent purification steps. After centrifuging out the cells, the supernatants were concentrated possibly by salt fractionation and/or pressure dialysis. The principal of salt fractionation is that proteins form hydrogen bonds with water molecules and these can be interferred with using salts causing the proteins to precipitate. Pressure dialysis is performed in a chamber where the protein solution is forced through a membrane under pressure. Various membranes allow the passage of different size molecules and the supernatants could thus be separated into molecular weight fractions and reassayed for biological activity. The fractions are also concentrated at the same time.

salt fractionation

pressure dialysis

gel filtration

isoelectric focusing

The fractions were then further fractionated principally by gel filtration and characterised as to molecular weight by SDS-PAGE or further purified by isoelectric focusing (separation by means of the isoelectric points of the proteins). In many instances, SDS-PAGE could not be used as the concentration of the cytokine was too low, even after concentration, and detection relied on the labour intensive biological assay. Even after all this work, the cytokines were not purified to homogeneity! A typical protocol is illustrated in Figure 6.1.

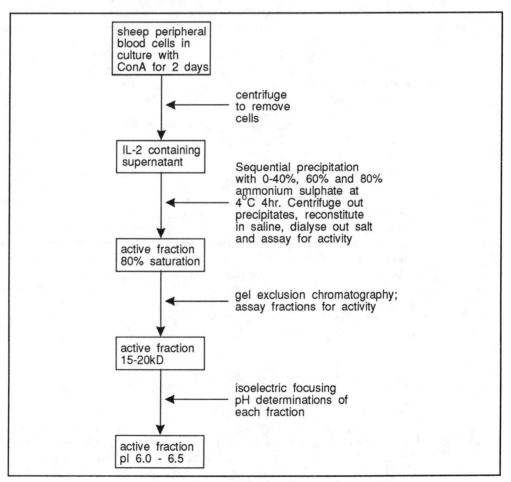

Figure 6.1 Protocol for the isolation of crude sheep IL-2 from sheep cell supernatants.

SAQ 6.1

Which of the following are correct/incorrect?

1) Treatment of T cells with anti-CD3 would result in the production of cytokines.

2) Protocols using salt fractionation, dialysis, gel exclusion and isoelectric focusing resulted in a pure active cytokine.

3) Treatment of T cells with mitogens results in production of IL-2.

4) Mixtures of cells from mother and offspring will induce IL-2 production.

5) Absence of activity in a supernatant does not necessarily indicate absence of mitogenic cytokines.

6.1.4 Monoclonal antibodies and gene cloning

monoclonal antibodies and gene cloning accelerated progress in cytokine technology

A couple of discoveries allowed more rapid progress in cytokine biotechnology. Cell lines were discovered in mouse and Man which selectively secreted, in much larger quantities, various individual cytokines. This allowed enough material for the production of monoclonal antibodies which then could be used for one step affinity purification of some of the cytokines. In one instance, 4 litres of supernatant was passed over a 1ml anti-Interleukin 2 column containing about 3mg antibody and this procedure removed all the Interleukin 2 from the supernatant. Additionally, the purified cytokine was recovered from the affinity column in 2ml acid buffer resulting in a 2000 times concentration! (Note that in some literature cytokines which influence the activities of lymph cells are referred to as lymphokines).

These purification procedures allowed at least partial sequencing of various cytokines and oligonucleotide probes based on these sequences were used to probe cDNA libraries. Other workers took a different approach and isolated mRNA from activated cells and translated this *in vitro* eg in *Xenopus* oocytes. In most cases, the translation product had to be assayed by the biological assay or using antibodies if they were available. Gene cloning methods were then used and the cDNAs expressed in *E. coli* and, in some cases, yeast or mammalian cells. Nowadays, it is the general practice to clone the cytokine cDNA once the biological activity is detected in a supernatant or cells and avoid all the purification steps referred to above. Whereas before very large numbers of activated cells had to be harvested to obtain enough material for cloning procedures, the polymerase chain reaction now allows amplification of the DNA prepared from the mRNA from just a few cells by reverse transcriptase. (Note the polymerase chain reaction is a process for producing multiple copies of nucleotide sequences. This technique is described in greater detail in the BIOTOL text 'Techniques for Engineering Genes').

polymerase chain reaction

6.2 Antigen specific factors

6.2.1 Major characteristics

antigen specific factors

We have already met such factors when we discussed antigen specific suppressor factors. Discussion of cytokines in the literature omits mention of antigen specific factors as they are not considered part of the cytokine family although they also qualify as cytokines. We will only refer to their major characteristics which are as follows. They are only produced by T cells during responses to antigen. The factors of molecular weight 40-50kD bind antigen and can be purified on an antigen affinity column. They also bear MHC determinants. For instance, murine helper T cell factors carry I-A determinants (see Section 4.2.2) and suppressor T cell factors I-J determinants (see Section 5.9.2). It is presently thought that these factors are soluble T cell receptors or parts of TCR.

6.3 General properties of cytokines

Π What do you think would be the principal characteristics of cytokines acting as messengers between cells. Before you read on, write down as many as you can.

6.3.1 Production and activation of cytokines

There are a number of features on the production and activities of cytokines which are common to all and you should be aware of these.

- they are produced during effector phases of both natural and acquired immunity;

- egulate development, activation and differentiation and effector functions of ne cells;

- re produced and act at extremely low concentrations;

- e often produced by a variety of cells and often act on many cell types. The ctivity is referred to as pleiotropism;

- e generally produced only for short periods of time;

- diate their actions by interacting with receptors on the target cells;

- or producers of cytokines are primed T cells although many other cell types cytokines as well.

es of action of cytokines

n. As we have just mentioned, many cytokines act on a variety of cell types h must express the receptor for that particular cytokine.

Many cytokines act in concert with each other. For instance, a cell may be release cytokine A which interacts with a receptor on another cell. This production of cytokine B or induces a receptor for a second cytokine.

autocrine, paracrine and endocrine action

Autocrine, paracrine and endocrine activities. A cytokine released by a cell may act on the producer cell to promote immune reactivity - this is autocrine action. The cytokine may act on neighbouring cells - this is paracrine action or on cells distant from the producer cell - this is endocrine action which is rare with cytokines.

6.3.3 Cytokines are antigen non-specific

there are
antigen and
cytokine
dependent
phases of the
response

Apart from those antigen specific factors we referred to earlier, all the cytokines are antigen non-specific. When T cells are activated by antigen, the antigen specific T cells release a number of different cytokines irrespective of the antigen specificity of the T cell. In other words, the same cytokines are used in humoral responses irrespective of the antigen provoking the response. The humoral immune response is initiated antigen specifically by processing and presentation of the antigen. Once activated, the T helper cells release a number of cytokines which then promote proliferation of B cells, clonal expansion and differentiation into plasma cells. In other words, the immune response requires antigen for its initiation and then becomes antigen independent but cytokine dependent for the subsequent stages of the response. This will become clearer later.

cytokines are
antigen
non-specific

6.3.4 Classification of cytokines

The cytokine field is a confusing one because each cytokine, in most cases, has been assigned many different properties and we do not know which of these properties are relevant *in vivo*. As we suggested earlier, many of the cytokines appear to synergise with others to promote an optimal response *in vitro* but these findings may have little relevance in the physiological situation. For the sake of you as the student, however, we have to attempt to classify the cytokines so you get some sort of reasonable picture of the field as a whole. We have done this in Figure 6.2 for cytokines we are reasonably sure about. Do not worry about the individual cytokines as the moment, we shall get to the details later. You will subsequently be able to use this figure as a summary sheet.

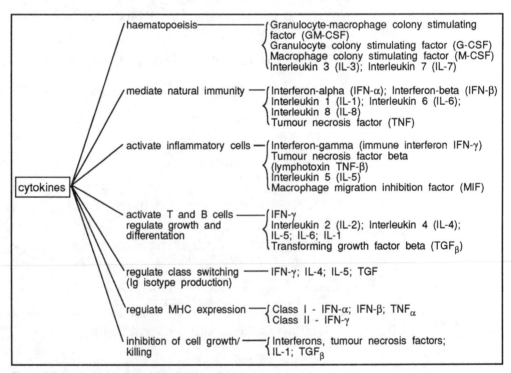

Figure 6.2 Involvement of cytokines in immunity.

6.3.5 Cytokine nomenclature

Interkeukins

By the end of the 1970s there had been reported in excess of one hundred biological activities in active supernatants and there was some confusion with regard to what to call each of them. It was realised that some of these activities were possibly due to the same cytokine and in 1979 the Interleukin system was introduced. The word interleukin means between leukocytes but this classification does not embrace all cytokines. The disadvantage with the Interleukin nomenclature is that the assignation does not provide information on the activity of the cytokine. For instance, one cytokine originally called T cell growth factor is now Interleukin 2. See what we mean? At last count we were up to IL-12.

SAQ 6.2

Match the items in the left column with those on the right using each item only once.

1) Paracrine	a) Receptor
2) Pleiotropic	b) Single cell
3) Interleukin 2	c) Neighbours
4) Autocrine	d) Concentrate cytokine
5) Antibodies	e) Many cell targets

6.4 The Interleukin 2 story

Discovery, structure and properties of Interleukin 2

We are now going to introduce you to the most well known cytokine Interleukin 2 as this can be your model for experimental approaches to identification and characterisation of individual cytokines and their receptors. Indeed, the methodology can to a great extent be applicable to any receptor-ligand system.

6.4.1 IL-2 promotes long term growth of T cells

Up until 1976 it was not possible to maintain long term growth of T cells. Then Morgan, Ruscetti and Gallo reported that T cells derived from the bone marrow could be maintained over months in culture if the culture medium was supplemented with supernatants derived from mitogen (phytohemagglutinin) -activated peripheral blood

T cell growth factor

cells. The active component was termed T cell growth factor (TCGF). At about the same time related activities from T cell supernatants were reported from other laboratories. Murine thymocytes at low cell concentration were found to be unresponsive to mitogen unless the culture medium was supplemented with a mitogen-activated spleen cell

co-stimulator

supernatant. The authors called the active component a co-stimulator. Other laboratories reported similar activities which they called thymocyte mitogenic factor and killer helper factor. These activities were eventually concluded to be due to the same T cell derived cytokine, this was named Interleukin 2 (IL-2).

Figure 6.3 gives an overview of the production and activities of IL-2. These features of IL-2 are covered in greater details in the following sections.

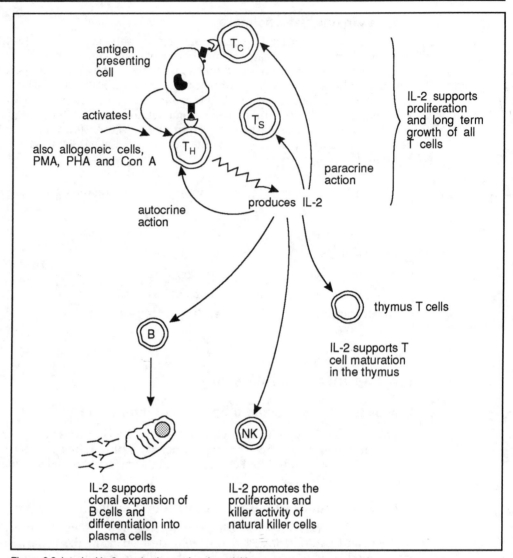

Figure 6.3 Interleukin 2, production and main activities.

6.4.2 IL-2 enables expansion of tumour specific cytotoxic T cells from solid tumours

IL-2 cloning of T cells

This cytokine held great promise since it was now possible to grow large numbers of T cells in culture. This facilitated cloning of T cells which eventually led to the characterisation of the TCR. Additionally, it was found that IL-2 could induce T-independent antibody responses in athymic mice and had great potential for the chemotherapy of cancer. Initial experiments demonstrated that if cell suspensions were prepared from solid tumours of mice and IL-2 was added to the culture, the resulting cells, when reinjected into the mice, mediated some regression of the metastatic growths.

∏   Can you explain this phenomenon before you read on.

tumour specific T cells

You can conclude that the tumour contained some tumour specific T cells which were expanded under the influence of the IL-2 and these killed off the tumour cells in the culture. These cells when reinjected into the mice attacked the tumour cells in the metastases. It has now been demonstrated that tumour specific cytotoxic T cells can be grown from solid tumour suspensions of tumour patients and these are being assessed for their effects *in vivo*.

6.4.3 Structure of IL-2

The IL-2 derived from human, rat and mouse T cells was subjected to separation protocols such as that shown in Figure 6.1 and found to have a molecular weight of about 15kD. Murine IL-2 has been recovered in a dimeric form of about 32kD.

IL-2 structure

A few years later, cDNAs were isolated for murine and human IL-2. The human cDNA consisted of an open reading frame coding for 153 amino acids, the first 20 constituting the signal peptide which is cleaved off to give a mature protein of 133 amino acids and a predicted molecular weight of 15.4kD. There is O-glycosylation of threonine at position 3 and a single disulphide bond crucial for biological activity between residues 58 and 105. The cDNAs were expressed in *E. coli* and recombinant IL-2 is available commercially. Some difficulties were encountered with the recombinant product due to aberrant disulphide bonding during refolding involving cys^{125} resulting in low activity of the IL-2. This can be avoided by site directed mutagenesis in which alanine has been inserted in human IL-2 and threonine in ovine IL-2. Human IL-2 uniquely acts on activated T cells of many other species whereas murine IL-2 only acts on rat and mouse cells. IL-2 from all species are more than 50% homologous to each other based on nucleotide sequencing. Murine IL-2 is rather unique in possessing a stretch of glutamine residues near the N-terminal end which is not found in any other species.

6.4.4 Production and biological activity

CD4[+] cells produce IL-2

IL2 is mainly produced by activated CD4[+] cells but also to a lesser extent by natural killer cells (large granular lymphocytes) and some CD8[+] T cells (we will discuss this more fully in Chapter 8). Although its main function is probably the clonal expansion of activated T cells, particularly promoting progression of T cells from G1 to S phase of the cell cycle, it appears to be a growth and differentiation factor for B cells and promotes development of lymphokine activated killer cells (LAK) which may have anti-tumour activity *in vivo*. At high concentrations, it also promotes growth and killer activity in NK cells and may act on macrophages. Finally, IL-2 is almost certainly involved in promoting growth of T cells in the thymus (see Figure 6.3).

SAQ 6.3	Which of the following are correct/incorrect

1) IL-2 is 153 amino acids long, is produced mainly by helper T cells and contains an intrachain disulphide bridge essential for biological activity.

2) At low concentration, IL-2 expands tumour specific T cells *in vitro* and promotes growth of B cells and NK cells.

3) In an antigen specific response IL-2 production is dependent upon MHC restricted cell interactions.

4) Activated CD4$^+$, NK and CD8$^+$ T cells produce IL-2.

5) T cells become IL-2 dependent at low cell concentrations.

6.4.5 Interleukin 2 assay

biological assay for IL-2

You will remember that we said that thymocytes would not respond to mitogen at low cell concentrations. This is due to a requirement for IL-2 and this has been used as the basis for a test for IL-2. A cytotoxic T cell line, CT6, was developed which was totally dependent on IL-2. The test is relatively simple. The cells are placed in wells of a tissue culture plate at a concentration of 10^4 to 10^5 cells per ml and re exposed to serial dilutions of IL-2 - containing samples. The control samples contain no IL-2. Radiolabelled thymidine is added over the last 4hr of a 24hr culture and the cells are harvested and assessed for uptake of the radiolabel. The control cells only show background levels of radioactivity whereas cells exposed to IL-2 exhibit uptake related to the levels of IL-2 in the medium thus indicating IL-2 dependent growth. Using IL-2 standards, concentrations of IL-2 are now defined in units of activity, one unit of activity being the amount of IL-2 required for half maximal stimulation of an aliquot of cells (Unit 2.4.2).

∏ We have already intimated that IL-2 acts through a receptor. Before you read on can you think how you could demonstrate the presence of receptors on IL-2 reactive cells in the laboratory. (You will be able to check your ideas with how this is achieved in the next section).

Following the development of antibodies to IL-2, this biological assay has been to a great extent superceded by an antibody based assay either a radioimmunoassay or the ELISA. In this latter case, enzymes linked to the antibody are used to measure the amounts of the antibody present (ELISAs are described in the BIOTOL text 'Techniques for Analysing Bioproducts').

6.4.6 A simple binding assay first IL-2 receptors on activated T cells

IL-2 binds receptors on activated T cells

A very simple technique first demonstrated that IL-2 receptors were present on cells responding to IL-2. Unprimed and activated T cells were incubated with a known amount of IL-2 (detected by the biological assay) in a culture dish for a short period. The culture fluid was then harvested and tested for the presence of IL-2. It was found that although the levels of IL-2 remained unchanged in cultures of unprimed T cells, the supernatant of the activated T cells were deficient in IL-2. This suggested that activated T cells express receptors for IL-2 whereas unprimed or resting T cells do not.

∏ If you have an antibody to the IL-2 receptor, what sort of tests would you do to establish the specificity of such an antibody?

Typical tests are shown in Figure 6.4. (Note that anti IL-2 receptor antibodies are referred to in this figure as anti-Tac. The explanation for this is given in Section 6.4.8).

6.4.7 A receptor binding assay

determination
of the number
of IL-2
receptors/cell

A receptor binding assay was then introduced to measure the amount of IL-2 bound per cell. In such an assay, various concentrations of radiolabelled IL-2 are incubated with activated cells and the cells are subsequently recovered by centrifugation and the supernatant containing the unbound IL-2 retained. The cells, with bound IL-2, are then centrifuged through a mixture of silicone and paraffin oil which removes any IL-2 not bound to the receptors on the cells. The centrifuge tubes are then frozen and the tips containing the cells sliced off and counted for radioactivity. Using various concentrations of IL-2 it was possible to determine the saturation point for a fixed number of cells ie when IL-2 was bound by all available receptors. It was then possible to determine the number of receptors per cell.

SAQ 6.4

In one such experiment, maximum binding of 4500dpm of ^{125}I-IL-2 was achieved using 450 000 cells. Assuming the specific activity of the radiolabelled IL-2 to be 1.7×10^6 dpm per picomol (10^{-12} mol) and Avagadros number to be 6×10^{23} molecules/mol, calculate the number of IL-2 receptors per cell.

6.4.8 Monoclonal antibodies to the receptor

Mice were immunised with a human T leukemic cell line which expressed high numbers of receptors for IL-2. The spleen cells were then removed from the mice and monoclonal antibodies prepared. One such antibody was designated anti-Tac indicating its specificity for activated T cells. Various tests using this antibody and illustrated in Figure 6.4 demonstrated that it was specific for the IL-2 receptor. This antibody, if immobilised on a column would retain a 55kD membrane component from

Tac antigen

lysates of activated T cells. This same component was also retained by an IL-2 affinity column and this membrane component was called the Tac antigen.

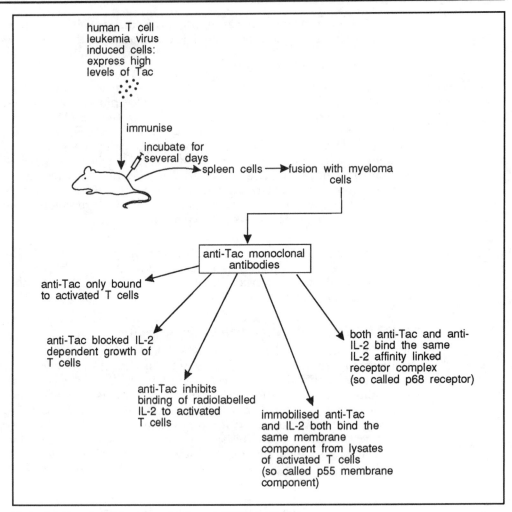

Figure 6.4 Experimental evidence for the specificity of anti-Tac.

6.4.9 High and low affinity IL-2 receptors

the cloned Tac antigen

Using an excess of radiolabelled anti-Tac in binding studies it was estimated there were about 50-60000 receptors per activated T cell. In parallel studies it was discovered that 5-10% of these were of high affinity and the remainder expressed low affinity for IL-2. The Tac antigen was cloned. The human cDNA encoding a 27kD peptide which with glycosylation resulted in a receptor chain of about 55kD. Since only one cDNA clone was identified, it was puzzling how there could be two types of receptor. The puzzle was solved by two classic experiments. The Tac cDNA was transfected into mouse fibroblasts and the cells were shown to express only low affinity receptors; if transfected into mutant mouse T cells which did not express the Tac antigen themselves, both high and low affinity receptors were expressed. The conclusion was that T cells constitutively expressed a second IL-2 receptor component which together with the Tac antigen formed the high affinity receptor.

6.4.10 The p55-p75 complex

an additional IL-2R chain

A few workers now independently demonstrated the presence of a second receptor chain. This was done using affinity linking techniques as shown in Figure 6.5.

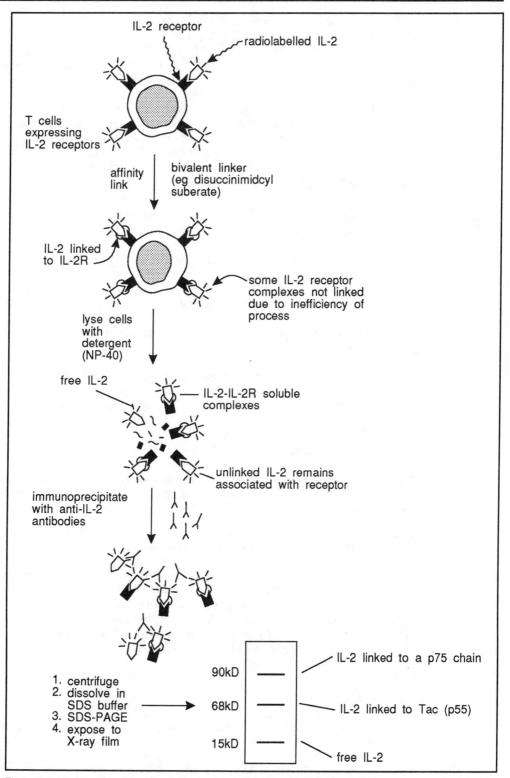

Figure 6.5 Identification of the p75 IL-2 receptor chain using affinity linking techniques (see text for details).

affinity linking

Briefly, activated T cells were incubated with radiolabelled IL-2. The cytokine was then covalently bound to the receptor using a linking agent such as disuccinimidyl suberate. The cells were then lysed using a detergent such as NP-40 and the IL-2-receptor complexes separated from the other membrane components by precipitation with anti-IL-2 antibodies. The complexes were recovered, dissolved in SDS-PAGE buffer, electrophoresed and the gels exposed to X-ray film. When picomolar levels of IL-2 were present during affinity linking, two bands resulted of 15kD and 90kD; in the presence of nanomolar levels of the cytokine an extra band of 68kD was present. The 15kD band represents free radiolabelled IL-2 which was not affinity linked but remained bound to the cells throughout the experiment. The 68kD band represented the Tac (p55) bound to IL-2 (15kD) and the 90kD band was a newly discovered IL-2 receptor component of about 75kD bound to IL-2.

| SAQ 6.5 |

Briefly explain the following observations made in the above experiment (refer to Figure 6.5).

1) Immunoprecipitation with anti-Tac resulted in 15kD and 68kD bands only.

2) Presence of anti-Tac during incubation of cells with radiolabelled IL-2 resulted in 15kD and 90kD bands.

3) Incubation of T cells in picomolar molar quantities of IL-2 resulted in 15kD and 90kD bands only.

4) Exposure of T cells to nanomolar quantities of IL-2 resulted in all 3 bands.

5) Presence of excess unlabelled IL-2 prevented appearance of any bands but Western blotting using a polyclonal anti-IL-2 resulted in 3 stained bands.

a low affinity receptor IL-2R$_\alpha$

The p55 (Tac antigen) was called the IL-2R$_\alpha$ chain and is the low affinity component (dissociation constant 10^{-8}mol l^{-1}). It has an extracellular domain of 219 amino acids, a transmembrane domain of 19 amino acids and short cytoplasmic tail of 13 amino acids. The rate of association with and dissociation from IL-2 of this chain has been estimated to be rapid (t$_{\frac{1}{2}}$ 10 seconds). A soluble form of this chain is found in supernatants and body fluids. It may be that one function of the α chain is to mop up excess IL-2 on the cell surface to prevent anergy (paralysis) of the cell.

an intermediate affinity receptor IL-2R$_\beta$

The p75 component is called the IL-2R$_\beta$ chain and the cDNA has also been cloned. There is a 214 amino acid extracellular domain, a 25 amino acid transmembrane domain and a large 286 amino acid cytoplasmic tail containing serine, acidic and proline-rich regions. This large tail is assumed to be involved in signal transduction of the high affinity complex. The IL-2R$_\beta$ binds IL-2 with intermediate affinity (dissociation constant 10^{-9} M), the rate of association of IL-2 being slow (t$_{\frac{1}{2}}$ of 45min) and dissociation very slow (t$_{\frac{1}{2}}$ of about 5hr).

∏ When the p55 and p75 components combine to form the IL-2 receptor, what do you expect the binding and dissociation rates to be compared to those of the individual chains?

high affinity complex

The high affinity receptor is composed of both chains bound to IL-2 and has a dissociation constant of 10^{-11} mol l^{-1}. The binding characteristics represent a hybrid of the two individual chains. There is rapid binding and slow dissociation ensuring that IL-2 is rapidly bound and retained by the receptor during internalisation. Various studies have suggested that amino acids 11-19 plus invariant aspartate at 20 of IL-2 bind the IL-2R$_\beta$ chain and amino acids 33 - 56 of IL-2 bind the IL-2R$_\alpha$ chain. It is likely that the whole complex is endocytosed during the activation of the T cells although there is evidence that only the IL-2R$_\beta$ chain needs to bind IL-2 to promote signal transduction, increased killer activity and proliferation in NK cells that do not express the α chain. We do not know which transmembrane signals are involved.

6.4.11 IL-2R associated proteins

Evidence suggests that the IL-2 receptor may be more complex. Additional proteins of 22kD, 35-40kD, a non IL-2 binding proteins of 75kD and a 95-105kD chain are all postulated to be associated with the two chain receptor. In addition, the IL-2R has also been found closely associated with both MHC Class I molecules and ICAM-1.

SAQ 6.6

Complete the table below indicating your answers by a +.

Structural feature	IL-2R$_\alpha$	IL-2R$_\beta$
Dissociation constant 10^{-9} mol l^{-1}		
p55		
Intermediate affinity		
Long cytoplasmic tail		
Soluble form		
Probable signal transduction		
Highest non peptide content		
Slow IL-2 association rate		
Binds IL-2		

6.5 Other cytokines involved in T/B activation

Let us now look at the structure and biological activities of other cytokines which are involved in the regulation of T and B cell responses. These include, in addition to IL-2, IL-1, IFN-γ, IL-4, IL-5, IL-6 and TGF-β. This section contains a lot of information, so it might be helpful for you to make a summary table as you read through this section.

6.5.1 Interleukin 1

Production of IL-1

Interleukin 1

Interleukin 1 was originally described as lymphocyte activation factor produced by macrophages activated by a variety of reagents such as endotoxin, adjuvants, PMA, urate and silica crystals, viral products and interactions with T cells. However, it is now known that many other cell types produce IL-1 although macrophages are the main producer. These include endothelial cells, dendritic cells, NK cells, astrocytes, kidney mesangial cells and even T and B cells! As well as the above reagents we also know that IL-1 production can be stimulated by a variety of cytokines (Figure 6.6).

Biological activities of IL-1

IL-1 stimulates
IL-2 production

IL-1 was originally described as being mitogenic for thymocytes and the original biological assay was based on this observation. It was found to stimulate the production of IL-2 by activated T cells and promote expression of IL-2R. In mouse, some T cells do not express receptors for IL-1 and appear to respond independently of IL-1. Expression of IL-2R is probably due to IL-6 which is released in response to IL-1. It is also involved in promoting chemotaxis of T and B cells. These are the principal effects of IL-1 in T/B activation processes.

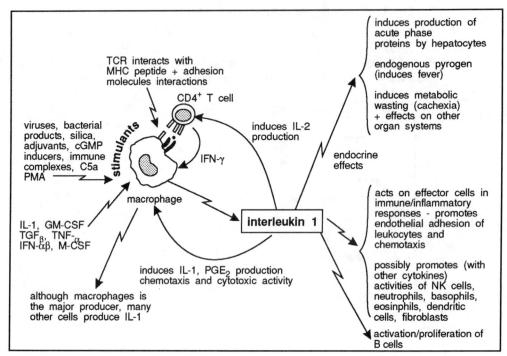

Figure 6.6 IL-1 endocrine activities.

IL-1 has
endocrine
activities

IL-1 is the one of the few cytokines which could have autocrine, paracrine and endocrine activities and exerts a wide variety of effects on a wide spectrum of cells and organ systems as shown in Figure 6.6. These include the ability to cause fever (endogenous pyrogen), induce synthesis of acute phase proteins and metabolic wasting (cachexia).

IL-1 structure

IL-1$_\alpha$ and IL-1$_\beta$

Biological activity of IL-1 resides in two distinct polypeptide species IL-1$_\alpha$ and IL-1$_\beta$. In Man, the latter is the major IL-1 and the main source is the macrophages. Both IL-1 species are about 17kD with isoelectric points of 5 (IL-1$_\alpha$) and 7 (IL-1$_\beta$) and exhibit somewhat less than 30% structural homology to each other. However, they both bind to the same receptor and appear to have identical biological activities.

IL-1 secretion
unusual!

Both species are synthesised as 33kD precursors which are proteolytically cleaved to give the mature products. Unlike other proteins destined for secretion, IL-1$_\alpha$ and IL-1$_\beta$ do not possess a hydrophobic leader sequence for transport of the polypeptide across the endoplasmic endothelium. Hence they are cytoplasmic proteins and the mode of secretion is not known. It is, however, thought that IL-1$_\alpha$ is processed to the 17kD form before being released whereas the IL-1$_\beta$ is produced in its inactive form.

IL-1 receptors

both IL-1 species bind the same receptor

IL-1 receptors are on most cell types examined. The receptor has been cloned and is an 80kD molecule with an extracellular domain of 310 amino acids, a transmembrane segment of 21 amino acids and a cytoplasmic portion of 217 amino acids. The receptor is a member of the Ig superfamily. Both IL-1 species bind to this receptor with high affinity. Another receptor of 60kD has recently been described in B cells.

IL-1 inhibitors

inhibitors for IL-1

There are naturally occurring inhibitors for IL-1. One is structurally homologous to IL-1 and binds to IL-1 receptors but is biologically inactive. It therefore competes with IL-1. Another is called uromodulin and is found in urine.

SAQ 6.7

Which of the following are properties of IL-1, IL-2 or both? Indicate by a +.

Property	IL-1	IL-2
1) Macrophage-T interaction induces?		
2) Bind receptors on T cells		
3) Two species of the cytokine		
4) Hydrophobic leader sequence		
5) Single receptor chain		
6) Clonal expansion of T cells		
7) Endocrine activity		
8) Receptors on majority of cells?		

6.5.2 Immune interferon- IFN-γ

Production and structure

two species of IFN-γ

IFN-γ is produced by activated CD4$^+$ and to a lesser extent CD8$^+$ and NK cells. In mice, IFN-γ is only produced by a subset of T helper cells, the TH1-like cells (we shall discuss these later). Original investigations using gel exclusion suggested a molecule weight of between 35 and 70kD. Human IFN-γ cDNA encodes a product of 166 amino acids including a signal sequence of 23 amino acids. The mature product (polypeptide backbone of 18kD) is found in two forms of pI 8.3 and 8.5 and molecular weights of 20kD and 25kD respectively after various degrees of glycosylation. They are also found as homodimers. The amino acid sequence is unrelated to those of the other interferons IFN-α and IFN-β and there is little sequence homology between species.

IFN-γ receptor

human IFN-γ receptor

A specific receptor has been identified which is distinct from the receptor for the other two interferons. The human IFN-γ receptor is a 90kD single chain glycoprotein and possesses a 229 amino acid extracellular domain, a 20 amino acid transmembrane region and a 223 amino acid cytoplasmic region which probably promotes intracellular signalling. Some workers report a much larger receptor on monocytes and other haematopoietic cells of about 140kD but this is thought to be due to extra glycosylation.

IFN-γ biological activities

stimulates IL-1 production

macrophage activating factor

One of the main activities of IFN-γ is as a macrophage activating factor which promotes the killing of intracellular pathogens. More about that later. More importantly, with regard to our present discussion, IFN-γ is the principal cytokine released by CD4$^+$ T cells when exposed to the correct MHC-peptide combination. This results in production of IL-1 which in turn activates the T cell to produce IL-2.

MHC II expression

IFN-γ also induces or enhances the expression of MHC Class I and II on target cells. It is a particularly important enhancer of MHC II in APCs such as macrophages and Langerhans cells as well as endothelial cells and some epithelial cells. Obviously this would promote increased antigen presentation to CD4$^+$ cells.

Some evidence suggests that IFN-γ also augments IL-2R expression on T cells and may act on proliferating B cells to promote differentiation to plasma cells. It induces Ig class switching to IgG2a in the mouse and actually inhibits switching to IgE production. It may also promote differentiation to effector cytotoxic T cells.

It also activates neutrophils and NK cells and promotes adhesion of blood borne CD4$^+$ T cells to endothelial cells by inducing expression of adhesion molecules on the latter. IFN-γ induces anti-viral defences in infected cells and is antiproliferative to many cells.

SAQ 6.8

Which of the following are true or false?

1) On being stimulated by MHC II-peptide on a macrophage, an antigen specific T cell releases IFN-γ which induces IL-1 release by the macrophage. This binds to a receptor on activated T cells and induces the release of IL-2.

2) IFN-γ and IL-2 are produced by all T cells and enhance the expression of MHC II on APCs.

3) IFN-γ is involved in the activation of inflammatory cells, activation of T/B cells and promotes IgG and IgE class switching.

4) IL-1$_\alpha$ and IL-1$_\beta$ are secreted as mature 17kD products.

5) IFN-γ exhibits paracrine activity.

6.5.3 Interleukin 4

IL-4

IL-4 is a 20kD glycoprotein produced mainly by activated CD4$^+$ T cells but also by some mast cells and bone marrow stromal cells.

In mice, IL-4 is a product of the TH2 CD4$^+$ T subset and has been implicated in the activation, proliferation and differentiation of B cells and as a growth factor for a subset of T lymphocytes (TH2). IL-4 promotes Ig class switching to IgG1 and, at high concentrations, to IgE. It also upregulates MHC II and an IgE receptor (CD23) on resting B cells.

growth factor for mast cells

It is also a growth factor for mast cells in synergy with IL-3. It is a macrophage activating factor like IFN-γ but is not as effective. It is often antagonistic to the effects of IFN-γ.

Human IL-4 cDNA encodes a 153 amino acid sequence of which 22 constitute the signal sequence. It has 2 possible N-glycosylation sites and six cysteine residues involved in disulphide bonding. It has about 50% homology with murine IL-4.

IL-4 receptors of about 60kD have been found on T and B cells, mast cells, fibroblasts and myeloid cells among others. There is also a truncated soluble form which may be a natural inhibitor of IL-4.

6.5.4 Interleukin 5

IL-5

proliferation and antibody production in murine B cells

IL-5 was originally described as a 45kD homodimer B cell growth factor II as it stimulated the proliferation and antibody production in murine B cells. Any similar effects on human B cells is controversial. It is known to increase IgA production in murine B cells but is not a class switch factor. It also augments high affinity IL-2R expression in resting and activated B cells.

eosinophil differentiation

Additional properties include a synergistic effect with IL-2 on the generation of cytokine (lymphokine) activated killer cells, promotion of eosinophil differentiation and activation of eosinophils.

6.5.5 Interleukin 6

IL-6

IL-6 has been known by a variety of names such as IFN-β2, B cell stimulation factor II. B cell differentiation factor and plasmacytoma growth factor among others. Human IL-6 may be a collection of cytokines of molecular weights 21-28kD depending on the degree of glycosylation. Macrophages are a major source of this cytokine but it is produced by T cells as well. In mice TH2 cells produce IL-6.

differentiation of B cells

Its main function in the humoral response appears to be in promoting differentiation of B cells into plasma cells. It may also play a role in T cell activation, possibly by enhancing expression of high affinity IL-2R.

It has many other reported diverse biological activities. It elicits the production of acute phase proteins by hepatocytes after being induced by IL-1 and TNF. It is involved in the differentiation of stem cells in the bone marrow, being synthesised by stromal cells and endothelial cells.

IL-6 receptor

The IL-6 receptor is present on many cell types. It is an 80kD glycoprotein consisting of an extracellular segment of 340 amino acids, a transmembrane region of 28 amino acids and a cytoplasmic segment of 82 amino acids. The N-terminal 90 residues is an Ig-like domain, making IL-6R a member of the Ig superfamily.

| **SAQ 6.9** | Assign each of the following statements/properties to IL-4, IL-5 or IL-6. Indicate your choice with a +. |

	IL-4	IL-5	IL-6

1) Induces class switching to IgG in mice

2) Induces acute phase proteins

3) Promotes production of IgG1 and IgE

4) 3 intrachain S-S bonds

5) B cell differentiation factor

6) Mast cell growth factor

7) Macrophage activating factor

8) Augments IL-2R in B cells

9) Natural inhibitor

6.5.6 Transforming growth factor beta

TGFβ

There is a family of about 6 transforming growth factor polypeptides of about 25kD which were originally described as inducing transformation of non-neoplastic cells in culture. They are made by macrophages, T and B cells and other cells and generally inhibit development and proliferation of cells in culture.

class switching to IgA

The one important property of importance to us in this chapter is in the ability of TGFβ to promote class switching to IgA production whilst at the same time suppressing production of other isotypes.

| **SAQ 6.10** | Before we construct the updated picture of the role of cytokines in T and B cell activation, take some time to work out the following problem.

Rabbits were injected with a crude sheep IL-2 supernatant preparation and the antiserum (anti-IL-2S) was used in biological assays and affinity linking experiments to partially characterise the sheep IL-2 receptor. Recombinant human IL-2 and a polyclonal antiserum to human IL-2 (anti-IL2H) was also available in the laboratory. Results of some of the experiments are shown in Figure 6.7. Which of the following conclusions can be supported from the data?

1) The anti-sheep IL2 antiserum is specific for the IL-2R as it blocks the biological activity of IL-2S.

2) Anti-IL-2H is not specific for the active site of IL-2H.

3) Based on the experimental evidence, IL-2H and IL-2S possess similar biological activity and high structural homology.

4) Sheep and human IL-2 bind to different sites on the sheep IL-2 receptor.

5) IL-2S is probably not pure sheep IL-2. |

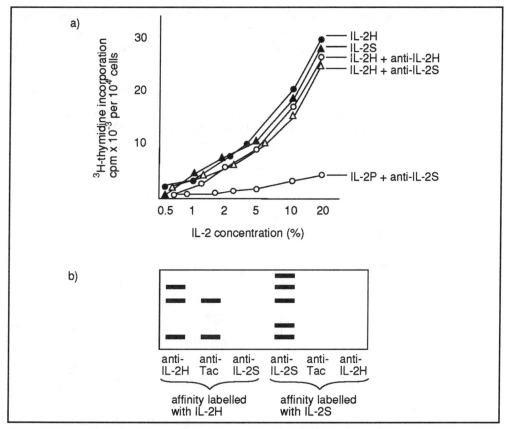

Figure 6.7 See SAQ 6.10. a) Biological assay: activated sheep cells were cultured at low cell concentrations with human or sheep IL-2 +/- the antibodies as indicated. Growth was assessed using the incorporation of ^{3}H-thymidine. b) Affinity linking: activated sheep cells were incubated with radiolabelled human or sheep IL-2 for 1hr at 4°C. The cells were then lysed with NP-40 and the affinity linked complexes immunoprecipitated with the antibodies shown followed by SDS-PAGE of the immunoprecipitates and autoradiography.

6.6 Cytokines and T cell activation

Now let us try to get a reasonable picture of what happens during T cell activation. We have to warn you there is a lot of speculation in what we are going to say.

⊓ Examine Figure 6.8. Read through it carefully and then use the following description to help you understand it.

6.6.1 Activation of the T cell - progression from G0 to G1 of the cell cycle

In a physiological response, the antigen presenting cell will process antigen and present peptides on the surface associated with MHC II. CD4$^+$ helper T cells with specificity for any of the peptides + MHC II will cross-link the ligands with TCR-CD3. Cell adhesion at this stage will be enhanced by low interactions between LFA-1 on the T cell and ICAM-1 on the APC and CD4 on the T cell binding to the monomorphic regions of the MHC II. We do not know the order of events but ICAM-1 expression on the APC increases along with increased avidity of LFA-1 - ICAM-1 interactions. All of these interactions result in production of IFN-γ by the T cell and APC binding of this results in what is called a co-stimulatory signal to the interacting cells. CD2 on the T cell now binds avidly with LFA-3 on the APC and this is followed by IL-1 production. IL-4 produced by the activated T cell may also be involved in an autocrine fashion in stimulation of the T cell perhaps by inducing IL-1 receptors. These stimulatory signals collectively induce the cell out of the resting (G0) phase into G1 of the cell cycle.

Remember that other activation signals may be involved since T cells also express CD26 and VLA antigens which interact with the extracellular matrix.

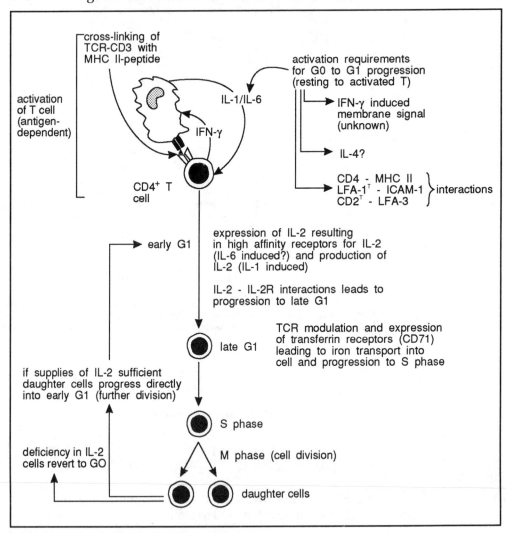

Figure 6.8 T cell activation and the role of lymphokines.

6.6.2 Cell cycling

IL-1 and IL-6   The activated T cells, on binding IL-1, express IL-2 mRNA and secrete this lymphokine. Another product of the APC (also produced by T cells), IL-6, is thought to induce high affinity IL-2R expression. The IL-2 binds IL-2R and the complex is internalised, promoting progression to late G1 of the cell cycle. As well as autocrine effects of IL-2, the secreted product will also exert paracrine effects on neighbouring T cells being activated on the same APC.

receptors for
transferrin

In late G1, the T cells express receptors for transferrin (CD71). All proliferating cells, not just T cells, show a requirement for iron and this is an essential requirement for cell cycling to proceed past last G1. Cells cultured in serum-less culture medium with no iron supplement, even with a plentiful supply of IL-2, fail to progress past G1. Although there has been some suggestion for cytokines in later stages of the cell cycle, you can assume that as long as IL-2 is plentiful and transferrin is available, the cells will progress through the cell cycle and the resulting two daughter cells will proceed direct into G1 for further rounds of proliferation.

6.6.3 IL-2 dependence and antigen availability

Although in the laboratory, you can remove the antigenic stimulus and maintain proliferation of the activated T cells with added IL-2 thus suggesting antigen independent, IL-2 dependent cell cycling, that is not the case *in vivo*. The presence of antigen is necessary for continued production of IL-2 and when supplies of antigen become depleted and IL-2 production becomes scarce, the cells stop cycling and revert to G0. The overall result of this process is clonal proliferation and production of memory helper T cells.

∏ Before you proceed, remind yourself of the suggested mechanisms for activation of B cells by T cells and FDC in the last chapter.

SAQ 6.11

Complete the following sentences (use words from the selection given below).

1) Activation of T cells and progression from [   ] to [   ] of the cell cycle involves cross-linking of [   ] - [   ] with antigenic [   ] bound to [   ], interactions between [   ] of the T cell and APC, production of [   ] by the activated T cell and [   ] by the APC.

2) [   ], secreted by APC, promotes the expression of [   ] IL-2R on the T cell. [   ], stimulated by the APC derived cytokine [   ], promotes the progression of the T cell through [   ] of the cell cycle when the T cell expresses receptors for [   ].

3) IL-2 production is dependent *in vivo* on the continued presentation of [   ] by [   ]. In the laboratory, T cell [   ] is antigen [   ] and IL-2 [   ].

Word List:

IL-6, IL-2, IL-1, transferrin, antigen, G0, G1, TCR, CD3, IL-2, APC, proliferation, peptides, MHC II, IL-1, cell adhesion molecules, independent, dependent, high affinity.

6.7 Cytokines and B cell activation

As for T cells we have a fairly confused picture of what is going on but it is improving. The B cell captures antigen and processes and presents peptides associated with MHC II. This particular B cell could be in the extra-follicular areas as part of a T cell-IDC complex or could be a memory B cell being exposed to antigen presented as iccosomes on follicular dendritic cells.

6.7.1 B cell activation - progression from G0 to G1 of the cell cycle

A memory T cell interacts with the B cell by weak interactions initially between LFA-1 and ICAM-1 on the B cell. This is followed by crosslinking TCR-CD3 with MHC II - peptide and CD4 binding to monomorphic MHC II. It has been postulated that this is followed by a membrane co-stimulatory signal perhaps mediated by a protein synthesised by the T cell. The B cell expresses the activation antigen B7 and the T cell secretes IL-4 and possibly IFN-γ. Both these cytokines may upregulate MHC II on the B cell increasing antigen presentation. Further avid interactions between CD2, CD28 and LFA-1 on the T cell and LFA-3, B7 and ICAM-1 on the B cell respectively pushes the B cell from G0 to G1 of the cell cycle.

activation antigen B7

6.7.2 Cell cycling and differentiation

IL-1, IL-2, IL-4 and IL-6

Early in G1 the B cell may produce IL-1 which stimulates IL-2 production by the T cell. This promotes proliferation of the T cell. The T cell also produces IL-4 which binds to B cell receptors and promotes progression to S phase and cell cycling. IL-2 may also be involved in cell cycling. As with T cells, B cells express transferrin receptors after IL-4 stimulation and take in iron for growth. Sometime during cell cycling some B cells express receptors for cytokines such as IL-6 which promote differentiation of the cell into a plasma cell and production of antibodies. Once again, IL-2 may act as a differentiation factor. In mice, IL-5 has been postulated to act at various stages of the response as shown in Figure 6.9.

∏   What is the overall result of B cell activation? (Examine Figure 6.9 carefully).

By examining Figure 6.9 you should be able to list 3 major consequences of B cell activation. They are T cell proliferation (clonal expansion), B cell clonal expansion and antibody production.

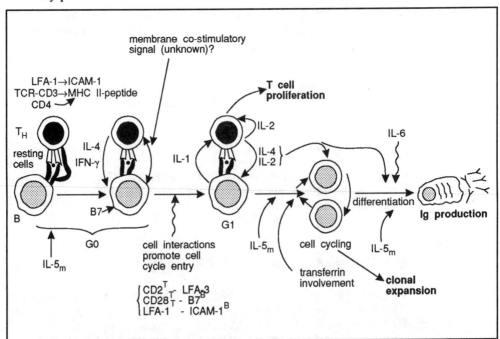

Figure 6.9 B cell activation, cell adhesion and lymphokines.

6.8 Ig class switching and production of isotypes

class switching   Some lymphokines have been shown to promote class switching in B cells to particular isotypes. IFN-γ induces class switching to IgG2a in mice and also inhibits switching to IgE production. IL-4 induces class switching to IgE and IgG1 although there is some doubt as to its requirement for IgG1 production. It has been shown to block switching of endotoxin-stimulated murine splenic B cells to IgG3 and IgG2b, instead promoting switching to IgG1 and IgE. Recent evidence suggests that IL-6, although promoting production of antibodies generally, may induce class switching to IgG1. Finally, TGFβ induces class switching to IgA and simultaneously blocks switching to production of other isotypes. IL-2 or IL-5 may modify this inhibition while at the same time enhancing the production of IgA.

Hence the overall result of B cell activation is the production of antibody and clonal expansion of B cells to memory cells. Do not forget, this also includes Ig class switching and somatic hypermutation to high affinity memory cells. Since B cells are potent APCs, B cell activation also results in expansion of the memory T cell pool.

SAQ 6.12	Match the items in the left column with those in the right column using each item only once.

1) LFA-1   a) LFA-3

2) IFN-γ   b) IL-4

3) IL-6   c) MHC-II-peptide

4) IgE/IgG1   d) TGFβ

5) CD2   e) B cell differentiation

6) TCR-CD3   f) IgG2a

7) Inhibits Ig synthesis   g) ICAM-1

6.9 Helper T cell subsets

two major subsets of CD4⁺ helper T cells   In mice, two major subsets of $CD4^+$ helper T cells have been identified. They are called TH1 and TH2 and have the capacity to produce different cytokines following activation. TH1 produce IFN-γ, IL-2 and TNFβ whereas TH2 produce IL-4, IL-5, IL-6 and IL-10. Both subsets appear to be able to produce other cytokines such as GM-CSF, IL-3 and TNFα to varying degrees.

This restriction in cytokine production means that each subset will be biased in the production of Ig isotypes. For instance, TH2 cells promote IgM, IgG3 and IgA production but have a bias towards production of IgG1 and IgE due to their ability to produce IL-4. TH1 cells are biased towards the production of IgG2a due to their production of IFN-γ although the cells also produce IgM and IgG3. However, TH1 cells can induce some IgG1 and IgE production if supplied with exogenous cytokines but are inhibited somewhat by producing IFN-γ which suppresses IgE production.

TH1 cells mediate cell mediated immune reactions, a so-called delayed hypersensitivity response (Chapter 8) due to being able to produce IFN-γ and TNFβ, whereas TH2 promote growth, differentiation and activation of mast cells and eosinophils due to the production of IL-3, IL-4 and IL-5 resulting in antibody mediated release of various mediators (Chapter 7).

The two subsets also appear to be mutually inhibitory and this may explain the ability of the immune system to select defined responses to certain pathogens. For instance, IFN-γ, produced by TH1, inhibits proliferation of TH2 and IL-4-mediated activity whereas TH2 produces a recently discovered cytokine IL-10 which inhibits production of IFN-γ and proliferation of TH1 (Figure 6.10).

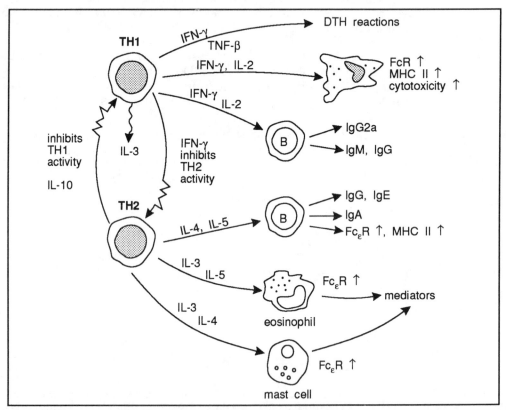

Figure 6.10 Biological activities of the murine T helper subsets TH1 and TH2.

6.10 Antibody synthesis

The synthesis of antibody molecules follows the cellular protocol for other secreted and membrane proteins. The only points we will mention are as follows. Heavy and light chains are synthesised separately within the endoplasmic reticulum (ER) but their synthesis is co-ordinated with surplus chains being degraded within the cell. Once the chains associate, the disulphide bonds are formed and carbohydrate is attached while the chains are in the ER. Sections of the ER are pinched off to form Golgi bodies containing antibody molecules; enzymes may further modify the attached carbohydrate and then the antibodies are secreted to the exterior of the cell by reverse pinocytosis.

SAQ 6.13

Complete the table indicating your choices with a +.

Feature	TH1	TH2
1) IFN-γ production		
2) IL-4 & IL-5 production		
3) DTH reactions		
4) Cytokines enhance MHC II expression		
5) Induce mast cell mediators		
6) IL-2		
7) IL-3		
8) Promote IgE production		
9) Activated by MHC II-peptide		
10) TNF-β production		

Summary and objectives

You have now completed a detailed study of cell interactions and will have considerable knowledge on experimental approaches, the importance of cell-cell contact promoted by adhesion molecules supplemented by the actions of different cytokines. We have only dealt with cytokines which have a role in T and B cell activation. As you will see in Figure 6.2 there are others which play major roles in hematopoiesis and natural and cell mediated immunity. We shall meet some of these in the last two chapters.

On the completion of this chapter you should be able to:

- devise protocols for isolating cytokines from supernatants;

- list the principle characteristics of cytokines;

- describe the production, structure and biological properties of interleukin 2;

- calculate the number of cytokine receptors on cells given experimental data;

- interpret experimental data using antibodies and affinity linking to define cytokine receptors;

- describe the structure and biological features of the IL-2 receptor components;

- list the major structural and biological properties of interleukin 1, immune interferon, interleukins 4, 5 and 6 and transforming growth factor-beta;

- describe the roles of cell adhesion and individual cytokines in activation, clonal expansion and differentiation of T and B cells;

- list the major distinguishing features of the murine helper T cell subsets.

The lymphatic system and the immune response

The lymphatic system and the immune response

7.1 Introduction

We are now going to develop your knowledge of physiological aspects of the immune system. We shall begin by looking at the structure and immune function of the lymphatic system, examine the development and major characteristics of other non-T/B cells which play important roles in the *in vivo* response and look at ways in which we believe cells move about the body. We shall then look at the kinetics of antibody production *in vivo* before we summarise the mechanisms operative in the physiological response. Finally we shall take a brief look at ways in which antibodies can cause tissue damage in a variety of clinical conditions.

7.2 The lymphatic system

7.2.1 General features

We learnt in Chapter 1 that the lymphatic system is part of the vascular system and mainly consists of lymphatic vessels filled with lymph (similar to serum but with less higher molecular weight proteins) and fixed and mobile lymphoid cells organised within lymphoid organs and tissues.

endothelium

A common structural feature of both the lymphatic and blood systems is the endothelium consisting of a layer of endothelial cells and an endothelial membrane. The endothelium is enclosed in supporting structures such as connective and muscular tissues in blood vessels and lymphatics. However, minute blood capillaries throughout the peripheral tissues consist solely of endothelial membrane with no supporting structures and these represent a gateway for migration of cells out of the blood into the tissues. These cells include T and B memory cells, monocytes and polymorphonuclear cells.

Some of the T cells (especially those which effect cell mediated immune reactions) may respond to environmental antigen which has crossed the external barriers and has been processed and presented by cells of the tissues. The T and B cells will finally make their way across the tissues and be picked up by the lymphatic system and will enter lymphoid organs such as lymph nodes. Antigen will also find its way either free or within APCs to the lymph nodes and interact with T and B cells resulting in activation, clonal expansion and antibody production. The T and B cell progeny and the freshly synthesised antibodies will be carried in the lymph from the node to the main lymphatic trunks. Many lymphatics carrying similar cells and products from different areas of the body finally merge into the thoracic duct which empties its contents into the blood stream via the subclavian vein in the neck.

recirculation of memory T and B cells

The lymphatic system, then, facilitates the recirculation of memory T and B cells from the blood to the tissues and lymphoid organs and then back to the blood. Other cells, which exit from the blood at similar sites, do not return to the blood but either die whilst dealing with pathogens in the tissues (polymorphs) or become fixed cells of the lymphoid system (monocytes/macrophages).

postcapillary
venules

As we shall find out later, T and B cells have additional gateways from the blood. There are specialised high walled endothelial cells in blood capillaries within lymph nodes called postcapillary venules. These allow exit of T and B cells only from the blood directly into lymph nodes without traversing the tissues.

transport of
nutrients

Additional functions of the lymphatic system are transport of nutrients from the intestine (the so-called lacteals) or tissues to the blood and the recovery of proteins lost from the blood. Figure 7.1 gives you some idea of the relationships between blood and lymph systems.

The important point for you to remember is that lymphoid tissue is strategically placed in most areas of the body to intercept antigen before it gains access to the blood stream ie to capture the antigen and keep it localised.

∏ To ensure that you know the main functions of the lymphatic system, make a list from the text.

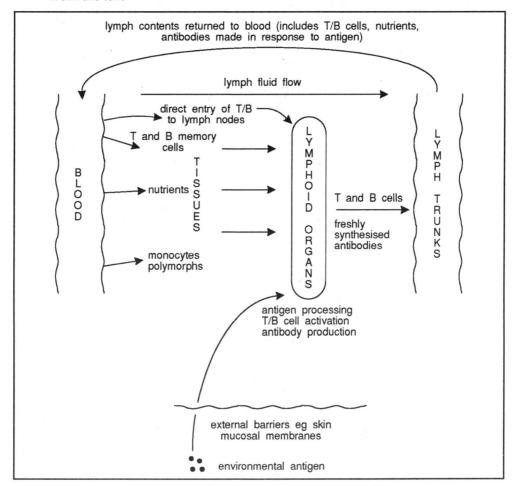

Figure 7.1 Functions of the lymphatic system.

7.2.2 Primary and secondary lymphoid tissue

Lymphoid organs and tissues are classified as being primary or central lymphoid organs or secondary or peripheral lymphoid organs. The primary lymphoid organs include the thymus and bone marrow, the foetal liver and the Bursa of Fabricius in chickens. These are the sites where the T and B cells develop from the stem cells, undergo DNA rearrangement and mature into immunocompetent cells. The secondary lymphoid organs include the lymph nodes, spleen and mucosally associated lymphoid tissues (MALT) and are the sites where immunocompetent T and B cells are activated by antigen resulting in humoral and cell mediated responses.

| **SAQ 7.1** | Four of the five items in each of the two sets below bear some common feature or function. Determine which is the odd man out in each of the two sets below. |

Set 1	Set 2
A) DNA rearrangement	A) Antibodies
B) MHC II-foreign peptide	B) Thymus
C) Foetal liver	C) Recirculatory pool
D) Thymic epithelium	D) Lymph nodes
E) CD1	E) Secondary lymphoid organ

7.2.3 The lymph node

Function of the lymph nodes

As we said earlier, many extracellular antigens or pathogens, after traversing the external barriers, become localised in the tissues. Here, they may be destroyed by wandering phagocytes. If this fate is avoided, they may be taken up by APCs or gain access to the vessels of the lymphatic system. Situated at intervals along these vessels are cellular filters called lymph nodes through which the lymph and its contents must traverse. These are able to trap particulate antigens such as bacteria and in the presence of specific antibodies, soluble antigens also. The lymph node has two main functions. Firstly, by trapping the antigen, it prevents access to the blood stream and the rest of the body. Secondly, the lymph node architecture provides the venue for T and B cells to be activated by the localised antigen which results in the production of antibodies (we shall deal with the production of effector T cells in Chapter 8).

Lymph node structure

sinus macrophages trap antigens

The main structural components of the lymph node are an external capsule from which radiates collagenous trabeculae into the body of the node. There is a network of fibres composed of reticulin throughout the node on which the cells are situated. Each node collects lymph and cells from a defined area of tissue called the drainage area using a number of afferent lymphatics which pour the lymph into the subcapsular sinus of the node. The lymph can flow down trabecular sinuses to medullary sinuses and finally into the hilum where the lymph exits from the node into efferent lymphatics which carries the lymph further upstream. The lymph may pass through a few nodes before it empties into the main trunks and finally the thoracic duct. The sinuses we refer to are lined with macrophages which are capable of phagocytosing and destroying antigen. The greater part of the antigen load is dealt with in this manner and is not involved in activation of acquired immune mechanisms.

cortex

The lymph node can be divided into three main areas, the outer cortex, the paracortex or deep cortex and the central medulla. Within the cortex you will find aggregates of cells called follicles, many of which possess a centre of actively dividing cells, the germinal centre. Follicles which contain germinal centres are called secondary follicles, those which do not have germinal centres are called primary follicles. Within the germinal centres you will find the follicular dendritic cells in close contact with B cells together with some T cells. This is where high affinity B cells are selected by antigen presented by follicular dendritic cells (FDCs). It is commonly referred to as a B cell area.

paracortex

post capillary venules

The predominant cell type in the paracortex is the T cell. This is where they can be found in close contact with interdigitating dendritic cells being primed by antigen (Section 5.8.3). The paracortex represents a direct entry point from the blood stream for 90% of all T and B cells found in the node. These cells apparently include T cells which have not yet been primed by antigen (Section 5.8.6). The capillaries allowing this selective migration are called post capillary venules and in most mammals contain high walled endothelial cells compared to narrower endothelial cells which allow migration of haemopoietic cells out of the blood. The significance of this structural difference is not clear.

medulla

plasma cells

The last area, the medulla is composed of the cellular medullary cords where we find plasma cells producing antibodies and the medullary sinuses lined with phagocytic macrophages. The medullary sinuses are continuous with the hilum which collects the lymph prior to its exit from the node via the efferent lymphatic.

The general structural features of the node are shown in Figure 7.2. Read the text on this figure carefully.

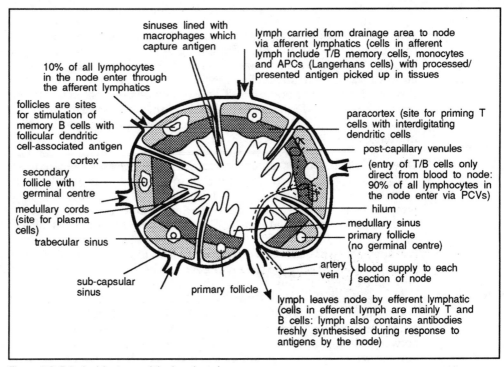

Figure 7.2 Principal features of the lymph node.

| **SAQ 7.2** | Which one of the following statements is correct? |

1) The major structural components of the lymph node are the capsule, trabeculae and a network of collagen fibres throughout the node matrix.

2) Within the cortex you will find follicles all of which contain germinal centres. These are sites for B cell-FDC interactions.

3) The paracortex is the T cell area of the node. This is the site where T cell and B cell priming may occur involving interactions with IDCs.

4) The medulla is made up of medullary cords and medullary sinuses. Prominent cells in these two areas are macrophages and plasma cells respectively.

5) The lymph nodes are primary lymphoid organs.

The humoral response in the lymph node

Soluble antigen and bacteria may gain access to the afferent lymphatics and be carried to the node where they are taken up by the macrophages lining the sinuses of the node. This antigen is destroyed and takes no part in induction of the adaptive response. Antigen is also endocytosed by cells such as Langerhans cells and monocytes as it crosses the external barriers into the tissues and these cells process the antigen, transport it to the node and then act as antigen presenting cells, particularly in the paracortex and extrafollicular areas, priming T cells and possibly stimulating memory T cells inducing clonal expansion and cytokine production.

recruitment phase

Within a few hours of antigen arriving in the node, we observe a massive increase in the numbers of T and B cells entering the node from the blood via the post-capillary venules. This phenomenon has been called the recruitment phase of the response and allows a wide spectrum of cells expressing various specificities to be exposed to antigen contained within the node. The T and B cells possessing specificity for the particular antigen are retained within the node where they become activated, resulting in clonal expansion of both helper T cells and B cells and production of antibodies under the influence of cytokines.

dissemination of the response

anamnestic response

Later in the response, when antibodies have been synthesised, antigen becomes concentrated on follicular dendritic cells in the cortex which function in selecting high affinity B cells into clonal expansion and differentiation into plasma cells. B lymphoblasts migrate across the cortex and paracortex and settle in the medullary cords as plasma cells producing copious amounts of antibodies which are secreted into the lymph and transported out of the node via the efferent lymphatic. Other B lymphoblasts will be carried out of the node in the efferent lymph and may eventually settle in another node, the spleen or even the bone marrow as plasma cells producing antibodies. We call this corporal distribution of B cells the dissemination of the response. Some B cells, of course, and T cells resulting from proliferation within the node, will become high affinity memory cells and will move from the follicles into the lymph and become part of the recirculatory pool. These cells will continuously travel between blood, the tissues and the lymph system until called upon to respond to antigen which has once again breached the external barriers somewhere in the body resulting in a secondary or anamnestic response.

SAQ 7.3

Which of the following is/are correct?

1) The afferent lymphatics are the main source of lymphocytes to the node.

2) Only a small percentage of the total antigen load is involved in inducing acquired immunity.

3) Cytokine production occurs in the paracortical and cortical areas of the node.

4) Cytokines and antibodies are found in efferent lymph.

5) Recruitment in a stimulated node increases the numbers of cells in efferent lymph.

7.2.4 The spleen

spleen acts as a filter of the blood stream

The spleen acts as a filter of the blood stream in much the same way as nodes filter the lymph and also provides the environment for activation of T and B cells as do lymph nodes. It should, however, be remembered that most of the antigens which find their way into the blood are removed by the liver.

red pulp

The basic structure of the spleen is of an encapsulated organ with similar supporting structures to that of the nodes. There are two areas within the spleen, the red pulp and the white pulp. The red pulp is considered to be the splenic equivalent of the medulla in that there are cords of cells interspersed with sinuses. Plasma cells are found there secreting antibodies. Like the medullary sinuses, those of the red pulp are lined with macrophages which remove antigen from the circulation. The red pulp is also concerned with the removal of damaged red cells from the circulation.

A cross section of the spleen will show that areas of white pulp are dispersed through the background red pulp. Indeed, the first histological description of the spleen was 'islands of white pulp in a sea of red pulp'. Each island of white pulp consists of a sheath of cells, principally T cells surrounding a central arteriole of the trabecular artery. The sheath is called the periarteriolar sheath (PAS) and is bounded by the marginal zone, an area rich in macrophages and antigen deposition. The PAS, then, is equivalent to the paracortex. Situated within the sheath are B cell rich follicles; these are equivalent to the cortex in the nodes (Figure 7.3).

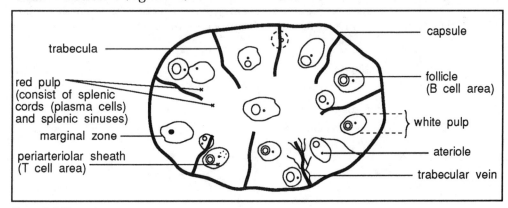

Figure 7.3 The spleen.

Lymphocytes can pass from the blood into the periarteriolar sheath where T cells can be activated by antigen presenting cells and B cells by either antigen + T helper cells or antigen-loaded follicular dendritic cells and T cell derived cytokines. The marginal zone is now thought to be a reservoir of persistent antigen which may also promote development of high affinity B memory cells as we learnt in Chapter 5. T and B cells can leave the white pulp via channels in the marginal zone and migrate into sinuses of the red pulp and thence into the blood stream.

ΙΙ Let us see what you remember about the structure of the lymph node. Close the text and draw a rough diagram of the cross-section of the node labelling as many details as you can remember. Check your diagram with Figure 7.2.

7.2.5 Mucosally associated lymphoid tissue

We also find nonencapsulated lymphoid tissue usually present as aggregates of B cells, dendritic cells and T cells, sometimes organised into follicles, below the mucosal layers in gastrointestinal, respiratory and urinogenital tracts. As an example you can find lymphoid follicles situated below the columnar epithelium of the ileum in an area called the laminar propria. Other well organised areas of follicles includes the tonsils and Peyers patches in the lower ileum. All of these mucosally associated lymphoid tissues (MALT) are prominent in their production of secretory IgA and, to a lesser extent, IgE which we talked about in Chapter 2. The gut associated lymphoid tissues are often referred to as GALT but are obviously part of MALT. IgA, produced by plasma cells within MALT, which does not link with the secretory peptide, will be taken up by lymphatic vessels draining this area and may finally appear in blood as serum IgA. IgA producing cells may also migrate through the lymph system and finally to the spleen.

There may be antigen-transport mechanisms across the mucosal layer. For instance, specialised epithelial cells lacking microvilli, called M cells, have been found in Peyers patches which possess many vacuoles that may be involved in the uptake and delivery of antigens from the gut lumen to the submucosal layer where contact would be made with antigen presenting cells, B cells and T cells. Many of the progeny of stimulated gut IgA B cells migrate out of the submucosa via the efferent lymphatics to the mesenteric lymph nodes and are then distributed to such sites as the small intestine, salivary and lacrimal glands and the mammary glands.

SAQ 7.4

To help you become familiar with the related structures of the spleen and lymph nodes complete the table inserting the name of the structure related to each of the activities in the left column. The same structure may appear more than once in the table.

Activity	Lymph nodes	Spleen
1) Antibody production		
2) T cell priming		
3) Location of IDCs		
4) B cell-FDC interaction		

7.3 Cells of the immune system

We have examined the characteristics and roles of T and B cells in great detail in past chapters and we now want to introduce you to some of the other cells which play important roles in the immune response.

7.3.1 Classification: humoral and formed elements

plasma

Blood consists of humoral (fluid) and formed elements. The humoral elements are contained in plasma, the fluid component of blood. This can be separated from the cellular elements by addition of an anti-coagulant such as heparin and then removing the cells by centrifugation. If plasma is allowed to clot and the clot removed, it is called serum and this is what we normally collect from an animal which has been immunised for preparation of specific antibodies. The blood is allowed to clot at room temperature and the clot is loosened with a spatula to free it from the collecting tube. The tube is then placed in a refrigerator overnight during which the clot will retract exuding the serum. The serum is then recovered and clarified by centrifugation.

granulocytes

mononuclear
cells

The blood cells can be divided initially into non-nucleated cells, the red cells and thrombocytes (platelets) and the nucleated cells (leukocytes) which are subdivided into granulocytes or polymorphonuclear (multi-lobed nucleus) cells and mononuclear cells. The granulocytes include the neutrophils, the eosinophils and basophils. These cells, once they have left the bone marrow, do not divide or return to the blood once they have left it. The mononuclear cells include the monocytes and lymphocytes and can divide in response to stimuli in the tissues. However, whereas the T and B cells can return to blood as part of the recirculatory pool, the monocytes are thought not to return once they have left but may reside in the tissues as tissue macrophages for long periods of time.

∏ Construct a classification diagram to help you distinguish between the different cell types and their characteristics. Check it against our Figure 7.4.

7.3.2 Haemopoiesis-development of blood cells

multipotential
stem cell

In the bone marrow and, earlier in development, in the yolk sac and the liver, stem cells arise which have the potential to develop into any type of blood cell. In other words, T cells, B cells, macrophages and neutrophils are all derived from the same precursor - the multipotential (pluriopotential) stem cell. We can now grow these cells in tissue culture and they can be stimulated to multiply and differentiate into colonies of various types of cells as Figure 7.5 shows, hence the name colony forming unit (CFU). By adding various cytokines to these cells we can influence their development. For instance, addition of granulocyte-macrophage colony stimulating factor (GM-CSF) will promote development of neutrophils and monocytes, granulocyte colony stimulating factor (G-CSF) will complete the differentiation to neutrophils and macrophage colony stimulating factor (M-CSF) complete differentiation to monocytes. Other cytokines we should mention are IL-3 and IL-7 both of which appear to exert their main function in haematopoiesis. IL-3 is known to promote development of the lymphoid lines but may also have an influence on differentiation of stem cells generally. IL-7 promotes development of immature B cells but has no effect on mature B cells. Erythropoietin promotes the development of red cells. The roles of many of the cytokines shown in Figure 7.5 are suggested but by no means proven.

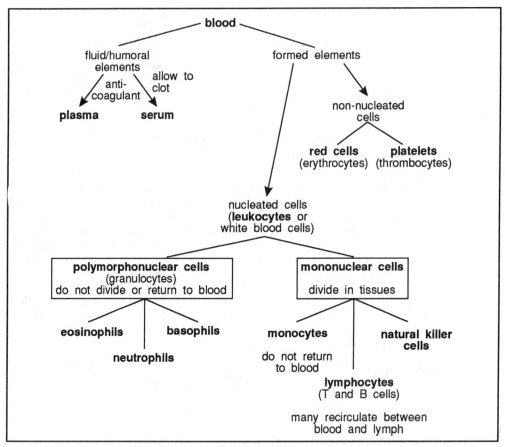

Figure 7.4 Cells of the immune response.

Π Examine Figure 7.5 carefully. Begin with bone marrow (on the left) and follow the development of each cell type. It might be helpful to write each sequence out separatel;y. In this way you will remember the details more clearly. For example, for the development of platelets, we might write the scheme:

$$\text{bone marrow} \longrightarrow \underset{\text{stem cell}}{\text{pluriopotential}} \overset{\text{IL-3}}{\longrightarrow} \text{CFU-GEMM} \longrightarrow \text{megakaryocyte} \longrightarrow \text{platelets}$$

We will not burden you with further details of the cytokines which appear to have a specific role in haemopoiesis. Notice that some of the cytokines involved in T and B cell activation may also be involved in development of blood cells. There is strong experimental evidence for cell interactions between the developing haemopoietic cells and cells called stromal cells during differentiation. Stromal cells are also known to produce a variety of cytokines which could influence haemopoiesis.

Once again, this area of research is in its infancy and there is a lot of work to be done to establish the true roles of cytokines in haemopoiesis.

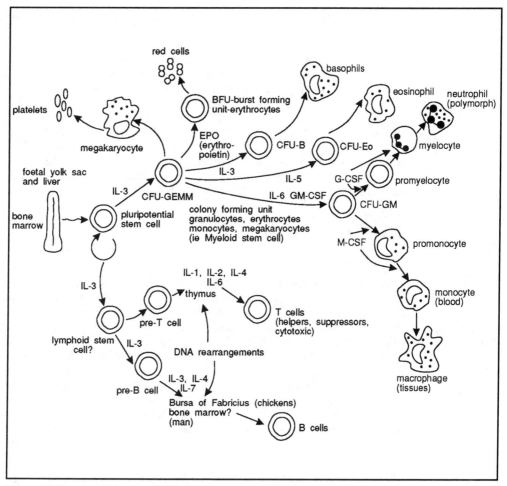

Figure 7.5 Haemopoiesis; the possible involvement of cytokines.

<table>
<tr><td>**SAQ 7.5**</td></tr>
</table>

SAQ 7.5	Match the items in the left column with those in the right column using each item only once.

1) B cells	A) G-CSF
2) Granulocytes	B) IL-3
3) T and B cells	C) IL-7
4) Macrophages	D) EPO
5) Red cells	E) M-CSF

7.3.3 Neutrophils

These comprise about 60-70% of the total white cell count and over 90% of the total granulocytes in the blood. There is a large store in the bone marrow as there is often a rapid response required of these cells when invasion by pathogens occurs. This is reflected by their short half life in the blood stream of about 12 hours. They are shortlived relative to macrophages, remaining in the tissues a few days. It is likely that a majority of these cells die while disposing of the invading pathogens.

The cells possess a multilobed nucleus and many granules or lysosomes. Primary granules contain myeloperoxidase, acid hydrolase and muramidase (lysozyme) enzymes. Secondary granules contain lactoferrin and lysozyme. Following ingestion of the antigens in a phagosome, lysosomes fuse with it to become a phagolysosome. The enzymes then destroy the antigens/pathogens. To facilitate capture of antigen, the cells express Fc_γ R111 receptors (CD16) for mainly IgG1 and IgG3 in Man and complement receptors CR1 (CD35), CR3 (CD11b/CD18) and CR4. CR1 binds C3b (and C4b), CR3 binds certain bacteria and a degradative product of C3b called iC3b. CR4 also binds iC3b. The most important one for you to remember is possession of CR1. Neutrophils also possess a receptor for C5a anaphylatoxin.

phagolysosome *(margin note)*

Π It would be a good idea for you to revise the use of these receptors in capture of antigens given in Chapter 2.

7.3.4 Basophils

(margin note) basophils contain histamine, heparin, ECF-A and produce PAF, SRS-A, thromboxanes and prostaglandins

These cells are present in very small numbers in blood (less than 0.2% of white cells) and appear to be the blood equivalent of mast cells in the tissues. They possess granules containing histamine, heparin, eosinophil chemotactic factor of anaphylaxis (ECF-A) and neutrophil chemotactic factor of anaphylaxis (NCF-A) among others which are released to the exterior of the cell by degranulation on stimulation of the cell. The cell also freshly synthesises other mediators on stimulation such as platelet activating factor (PAF), slow reacting substance of anaphylaxis (SRS-A, its effects are now known to be due to substances called leukotrienes), thromboxanes and prostaglandins. These cells possess the same complement and IgG receptors as neutrophils but, in addition, a high affinity receptor ($Fc_\varepsilon R1$) for the IgE made by plasma cells in the submucosa and other sites such as the skin. Mast cells are now known to release a wide spectrum of cytokines including IL-3, 4 and 5 which were once thought to be produced exclusively by T cells.

Basophils and mast cells are activated by crosslinking of the IgE receptors by antigens and any other reagents that can have the same effect such as anti-IgE.

You will remember that these cells also express receptors for C3a and C5a derived from complement activation and binding of these also induces degranulation.

Π Activation results in the release of many factors especially those referred to above. Before you read on, make a list of these and include their roles. You can check your table in the following two paragraphs.

Once an individual has become primed by antigen at a local site, subsequent exposure to antigen will result in crosslinking of IgE on mast cells and degranulation. NCF-A and ECF-A will diffuse towards the nearest venule and attract neutrophils and eosinophils from the blood stream and these will move up the chemotactic factor concentration gradient where the antigen is situated. The neutrophils will then phagocytose the antigen.

Histamine released by the mast cells will, by promoting vascular permeability (vasodilation) increase fluid passage from blood to tissues. This will include IgG and complement. You will remember that these are necessary for opsonisation of the antigen leading to phagocytosis. Activation of local macrophages will lead to production of other mediators which reinforce these activities. Included among these is IL-8 which is chemotactic for both neutrophils and basophils.

7.3.5 Eosinophils

These cells comprise up to 5% of the total white cells in blood. Eosinophils express the IgG and complement receptors found on neutrophils but in addition a low affinity receptor for IgE ($Fc_\varepsilon R11$, CD23). Although they are minor phagocytes, they appear to operate by a degranulation mechanism most of the time. Eosinophils are thought to be the principal cells involved in the disposal of worms or helminths which are many times bigger than they are. On being attracted to a site of worm infestation by ECF-A, the cells bind to the helminths coated with IgG or IgE. This is followed by degranulation and one of the products, major basic protein, is toxic to the helminths.

major basic
protein

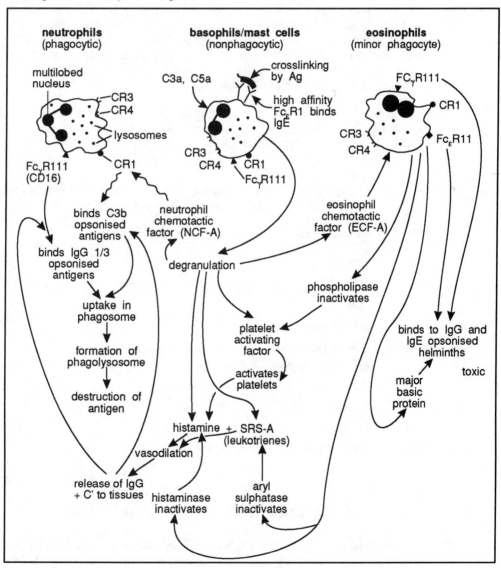

Figure 7.6 Neutrophils, basophils/mast cells and eosinophils.

The eosinophils also appear to down-regulate the activities of the mast cell by secreting aryl sulphatase, histaminase and phospholipase which inactivate the mast cell products SRS-A, histamine and PAF respectively.

The characteristics of neutrophils, basophils/mast cells and eosinophils are summarised in Figure 7.6.

7.3.6 Natural killer cells

These are a subset of lymphocytes which represent 5-10% of the total blood lymphocytes. They are large cells, alternatively called large granular lymphocytes, and possess cytoplasmic granules. They are neither T nor B cells. They do not mature in the thymus, are non-MHC restricted and do not undergo DNA rearrangements. They do express CD2 and $Fc_\gamma R111$ (CD16) but lack CD3. The major markers on these cells are CD56 and CD57. However, NK cells do have the CD3 zeta chain expressed as a homodimer associated with CD16. NK cells also express the $IL2R\beta$ chain and can be stimulated to proliferate with high concentrations of IL-2. They are also responsive to the cytokines $IFN-\gamma$, TNF and IL-2 which promote cytolytic activity of NK cells and produce low levels of IL-2 and some IL-3.

NK cells appear to have some cytolytic activity against tumour cells and against cells infected with viruses such as influenza virus and herpes virus. However, their role in the immune response is not known. You will get more information on these cells in Chapter 8.

SAQ 7.6

Complete the table indicating with a + the comparative characteristics of neutrophils, basophils/mast cells and eosinophils.

Characteristic	Neutrophil	Basophil/ Mast cell	Eosinophil
1) $Fc_\gamma RIII$			
2) Histaminase production			
3) Binds C3b			
4) High affinity $Fc_\varepsilon R$			
5) Vasodilation			
6) Low affinity $Fc_\varepsilon R$			
7) Granules			
8) Release of PAF			
9) Nonphagocytic			
10) Binds IgG opsonised antigens			

7.4 Cell migration and adhesion molecules

You have already been introduced to the idea that T and B cells can enter through postcapillary venules direct into lymph nodes whereas other haemopoietic cells cannot. Many experiments have been done that demonstrate selective migration particularly of T cells to certain tissues. Let us describe a typical experiment.

7.4.1 Demonstration of selective migration

Lymphocytes were collected from intestinal (mesenteric) lymph nodes by insertion of a plastic cannula into the efferent lymphatic vessel and from the prescapular lymph node (drains peripheral tissues) in a similar manner. The cells were incubated in a solution of radio-labelled chromate; the radio-label gets taken up by the cytoplasm of the cells. The radio-labelled mesenteric and prescapular cells were reinfused into the animal and the efferent lymph of mesenteric and prescapular lymph nodes was examined for the presence of radiolabelled cells. The results showed that the cells from the gut preferentially returned to the gut and the cells from the periphery returned to peripheral lymph nodes. In other words, there was selective migration of the lymphocytes.

Methods were developed subsequently which allowed the isolation of high endothelial cells (HEV) from various sites such as the postcapillary venules in peripheral lymph nodes and Peyer's patches in the gut. The HEV could be placed in tissue culture dishes and it could be demonstrated that lymphocytes could bind to such tissues. It was postulated that migration across endothelial membranes was dictated by receptors on lymphocytes interacting with ligands on the endothelial cells.

7.4.2 Homing receptors

The first monoclonal antibody prepared against such cells was called MEL-14 and this was found to inhibit the binding of mouse lymphocytes to HEV of peripheral lymph nodes but not to those prepared from Peyer's patches. The MEL-14 was found to bind a 80-90kD molecule called $gp90^{MEL}$ on lymphocytes and a 100kD molecule on neutrophils. It has been suggested that $gp90^{MEL}$ may be the T and B cell surface molecule (homing receptor) which dictates the passage of these cells only across HEV of peripheral lymph nodes by interacting with a molecule called PN-Ad (Section 7.4.3). The human equivalent of $gp90^{MEL}$ is LAM-1 (leukocyte adhesion molecule) and both have now been renamed L-selectin.

The structure of L-selectin has been worked out and consists of 3 regions. There is an outer lectin domain of about 120 amino acids adjacent to a short domain with homology to epidermal growth factor and a membrane proximal region consisting of 62 amino acid repeat structures showing homology to a complement binding protein. Other members of the selectin family include ELAM-1 expressed on endothelial cells and which binds to neutrophils (and possibly some memory T cells) and CD62 (PADGEM, GMP-140) also expressed on endothelial cells and which binds neutrophils and monocytes.

selectin

Another homing receptor molecule is CD44 of 85-95kD which has a broad distribution among haemopoietic and non-haemopoietic cells. We do not know the endothelial ligand for CD44 but it may fulfil a general function for migration of lymphocytes across all types of HEV.

SAQ 7.7

In some experiments, supernatants of rat lymphocyte subpopulations were found to contain soluble homing receptors. They were called HEBF$_{LN}$ (high endothelial binding factor) and HEBF$_{pp}$. Monoclonal antibodies A11 and 1B2 were prepared with specificity for one or the other of the homing receptors. Figure 7.7 shows the results of two experiments using these antibodies to identify rat T cells which selectively migrate to peripheral lymph nodes (cervical lymph node) or Peyer's patches.

Which of the following statements are reasonable conclusions based on the experimental evidence?

1) HEBF$_{LN}$ may be a homing receptor for peripheral lymph nodes and spleen.

2) A higher proportion of cells express homing receptors for Peyer's patch HEV than they do for cervical lymph node HEV.

3) The number of TDL binding to cervical lymph node or Peyer's patch HEV after treatment with anti-Ig antibodies probably represents the total cells in the TDL expressing HEBF$_{LN}$ or HEBF$_{pp}$ respectively.

4) Structures recognised by A11 and 1B2 do not affect entry to the periarteriolar sheath of the spleen.

5) Antibodies A11 and 1B2 are specific for HEBF$_{LN}$ and HEBF$_{pp}$ respectively.

7.4.3 Vascular addressins

Antibodies have been used to identify two molecules on endothelial cells which appear to be ligands for some of the homing receptors referred to. Antibody MECA-79 identifies a 92-110kD molecule called PN-Ad (peripheral lymph nodes-addressin) found on HEV of peripheral lymph nodes in Man and mouse which is bound by L-selectin. A second antibody MECA-89 identifies a 58-66kD molecule called M-Ad (mucosal - addressin) found on HEV of Peyer's patches and the mesenteric lymph nodes. The receptor for this ligand is not yet known but could be CD44, VLA-4 (Man) or LPAM-1 (mouse) as suggested in Figure 7.8.

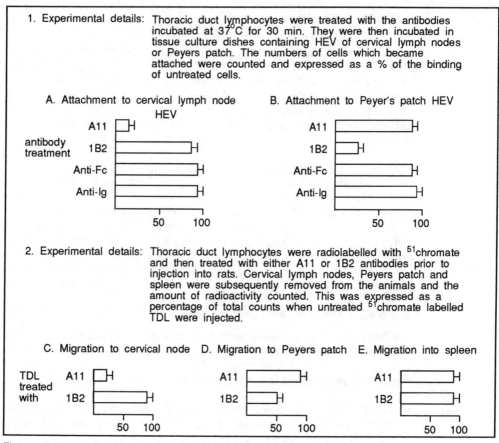

Figure 7.7 Rat homing receptors (see SAQ 7.7).

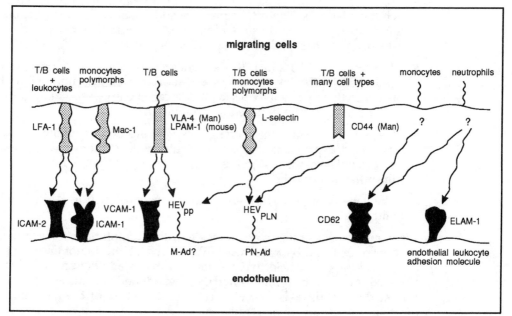

Figure 7.8 Cell adhesion molecules.

7.4.4 Integrins in cell migration

integrins   Another family of adhesion molecules are called integrins which are all composed of an
α and β chain. These include some of the VLA antigens which we met in Chapter 5. One
of these, VLA-4 (murine equivalent is LPAM-1), consisting of CDw49d/CD29, you will
remember is found on T cells and interacts with VCAM-1 on APCs. VCAM-1 is also
expressed on endothelial cells of Peyer's patches; so VLA-4 may dictate the migration
of lymphocytes across the HEV at these sites. You will remember that VLA-4 also
interacts with the extracellular matrix facilitating movement of lymphocytes across
tissues.

Another member of the integrin family is LFA-1, found on most leukocytes, which is
involved in cell-cell interactions (Chapter 5) but may also play a role in cell migration.
The receptors for LFA-1, as you know, are ICAM-1 (CD54) and ICAM-2 which are
present on endothelium or can be induced by cytokines. ICAM-1 can be induced on
endothelial cells by the cytokines IL-1 and TNF and may promote the enhanced
migration of cells such as lymphocytes and neutrophils across endothelia to
inflammatory sites since these cytokines are produced by activated macrophages.
Another integrin, Mac-1 (CD11b/CD18), found on monocytes and polymorphs, also
interacts with ICAM-1 and may promote the exit of these cells from the blood.

7.4.5 Possible roles of cell adhesion and cytokines in diapedesis

We have tried to summarise the main features of cell adhesion molecules in Figures 7.8
and 7.9. Although our knowledge is somewhat scant in this area we think you should
now have a reasonable picture of the role of cell adhesion and cytokines in cellular
migration. We suggest that the events may unfold as follows:

- Activation mechanisms involving T cells, macrophages, mast cells and complement
  activation results in an inflammatory reaction and production of (i) chemotactic
  factors and (ii) cytokines.

- IL-1 and TNF (lymphotoxin) secreted by activated macrophages activate the local
  endothelial cells. These cells produce additional chemotactic factors such as IL-8
  and up-regulate adhesion molecules for T and B cells and/or leukocytes
  (neutrophils, basophils, eosinophils or monocytes).

- The targeted cells attach to the endothelium via their homing receptors to the
  endothelial adhesion molecules and move across the endothelium by squeezing
  between the endothelial cells (a process called diapedesis) being constantly guided
  by and in contact with the adhesion molecules.

- Once in the tissues, the homing receptors are downregulated. There may be an
  increased expression of VLA molecules which interact with molecules of the
  extracellular matrix and guide the cell movement up the concentration gradient to
  the inflammatory site.

- The cells now perform their effector functions such as disposing of opsonised
  antigen.

We would expect similar sorts of events within lymph nodes which would effect the
movement of T and B cells across HEV. We, however, do not know at this stage how
certain cell types are selectively attracted across endothelium based on our present
knowledge of homing receptors and their ligands and cytokines. We shall have to await
further developments before we have a clear picture of the mechanisms governing
directional migration of individual cells.

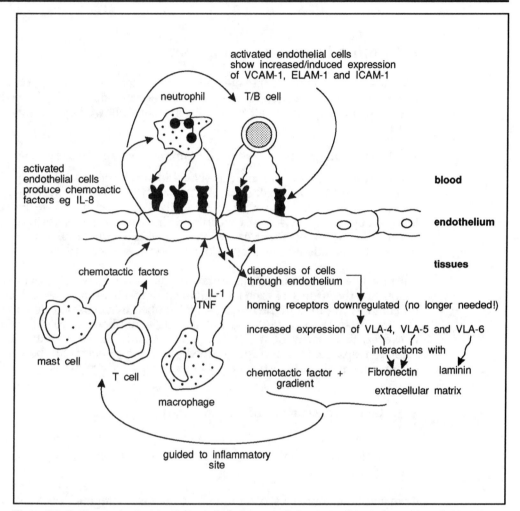

Figure 7.9 Guidance of cells through the endothelium to an inflammatory site (suggested mechanisms).

SAQ 7.8

Which one of the following statements is correct?

1) Mac-1, expressed on monocytes, selectively promotes the exit of these cells from the blood by interaction with ICAM-1.

2) LFA-1 and VLA-4 are structurally related.

3) The binding of leukocytes to endothelium and subsequent diapedesis is cytokine independent.

4) CD44 is expressed on many cell types and this, together with VLA-4 bind the vascular addressin M-Ad on endothelial cells.

5) The structure identified by MEL-14 is structurally related to ELAM-1 and CD62 both of which selectively promote the migration of neutrophils across the endothelium.

7.5 Antibody production *in vivo*

We have now examined the sites where cell interactions occur and antibodies are produced. We will now complete our discussions on antibody production by briefly describing the response kinetics observed in serum followed by a summary of the major features of the physiological response.

7.5.1 Major differences between primary and secondary responses *in vivo*

Let us imagine that we immunise a rabbit with an antigen and then withdraw samples of serum daily over the next two weeks. Antibody levels will be as shown in Figure 7.10. There is an initial lag phase when no antibodies are detected in the serum. This is followed by about day 4 with a logarithmic increase in the serum antibody levels (we call this the 'titre'). By about days 6-8 the levels of serum antibodies have levelled off and is subsequently followed by a slow decline in antibody titres.

The first detectable antibodies in serum are generally IgM which is of poor quality. This means that the affinity of each individual binding site of the IgM molecule is low and remains low throughout the response. IgG antibodies can be detected about a day after the first production of IgM and initially, these antibodies are also of poor quality. However, the affinity of IgG antibodies gradually increases with time during the primary response. IgM and IgG antibodies are the principal antibodies produced by peripheral lymph nodes. The kinetics of antibody production is shown in Figure 7.10.

∏   Can you explain the increase in affinity of IgG?

You should have been able to recall the mechanism for the change in affinity of IgG. If not, re-read Section 3.6.5.

If we inject the animal a second time with the same antigen 3 weeks after the primary injection, we observe dramatic changes in the kinetics of IgG production. There is a much shorter time lag between immunisation and detection of IgG, the amount of IgG in serum is vastly increased and the antibodies are of much higher average affinity. In contrast, the kinetics of IgM is the same as in the primary response. The production of high quality IgG is the aim of many immunisation programmes. The increase in quality of IgG is called the maturation of the immune response and involves somatic hypermutation events described in Chapter 3.

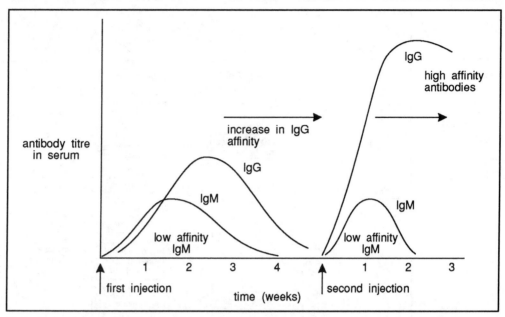

Figure 7.10 Serum antibodies during the physiological response. Note the change in the affinity of IgG as well as the titre levels.

<table>
<tr><td>

SAQ 7.9

</td><td>

Complete the table indicating possession of feature by a +.

</td></tr>
</table>

Feature	IgG	IgM
1) Activates complement		
2) Somatic hypermutation		
3) First detectable antibodies		
4) Primary/secondary response differs		
5) Protective antibodies		
6) High affinity		
7) B cell receptor		

7.5.2 Major features of the physiological response

We are making this very brief since by this time you should be able to make many connections between the different subjects we have discussed to come up with a fairly complete picture of what goes on in the individual when he/she is invaded by a pathogen. We have summarised these events in Figure 7.11.

∏ You should redraw Figure 7.11 onto a large piece of paper filling in details as you know them at each stage depicted in the figure. Then pin the sheet up as a remainder of the overall physiological response to infection.

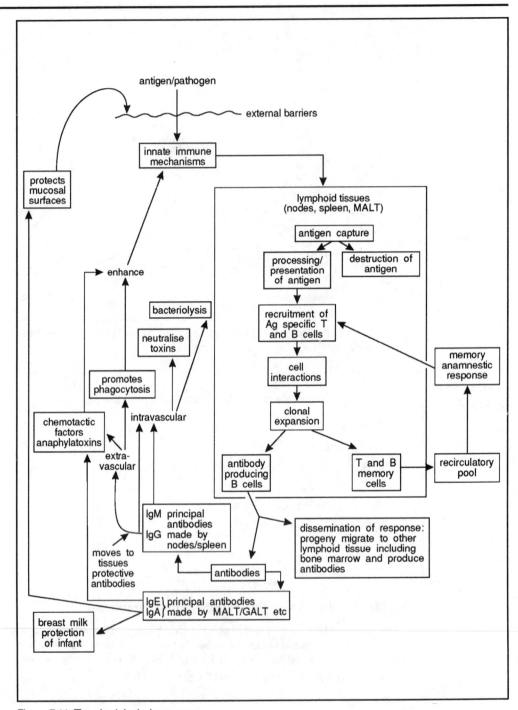

Figure 7.11 The physiological response.

7.6 Diseases caused by antibodies - the hypersensitivity reactions

hypersensitivity reactions

It has been emphasised that once an individual has been primed by antigen, subsequent responses are much larger and more effective due mainly to the improved quality and larger amounts of antibodies produced. In some instances, however, the response becomes harmful resulting in tissue damage and the responses are then called hypersensitivity reactions as the individual appears extra sensitive to the antigen. These often occur when antigen is present in large amounts or when antigen persists following infection. In some instances, certain individuals appear to be genetically predisposed to these conditions.

7.6.1 Classification of hypersensitivity

There are four main types of hypersensitivity I-IV. Types I-III (also called immediate hypersensitivity) are mediated by antibodies. These are Type I anaphylactic, Type II cytotoxic and Type III complex mediated. Hypersensitivity Type IV is also called delayed type hypersensitivity and is mediated by T cells. We shall deal with this in Chapter 8. We are not going to examine these conditions in great detail - we simply want to emphasis the damage that antibody can do in some individuals and under the right conditions.

7.6.2 Hypersensitivity Type I: anaphylactic reactions

The anaphylactic response

allergens

Antigens which induces this condition are called allergens and include insect venoms such as bee stings, pollens, animal danders (skin flakes), faeces from house dust mites which reside in our carpets and bedding, many different food allergens and drugs.

Once primed to the allergen, hypersensitive (atopic) individuals respond with a variety of symptoms to a second challenge. Allergens which make skin contact may induce skin rashes and raised, soft lumps (hives, urticaria) which may progress to vesiculation or blisters (atopic eczema). Allergen contact in the eyes leads to conjunctivitis, in the respiratory tract to rhinitis (runny nose) as is seen in hay fever or asthma, in the gastrointestinal tract to food allergens, cramps, nausea and vomiting.

A more dramatic and life threatening condition is induced if the allergen enters the blood stream. This is called systemic anaphylaxis and can lead to asphyxia, circulatory shock, collapse, unconsciousness and death. Injection of penicillin into a person hypersensitive to the drug can lead to death within 30 minutes (hence the term immediate hypersensitivity).

Priming and degranulation

preformed mediators

On first contact with the allergen, IgE is produced by B cells localised in skin and submucosa and this is taken up by tissue mast cells bearing high affinity receptors for IgE. Subsequent exposure of the individual to the allergen results in crosslinking of the surface IgE on the mast cells resulting in degranulation as we talked about earlier. This results in release of preformed mediators such as NCF-A and ECF-A which attract neutrophils and eosinophils to the site, heparin, an anticoagulant and histamine.

Histamine

Histamine acts locally to induce retraction of endothelial cells, in effect opening up capillaries (vasodilation) allowing fluid loss to the tissues. The fluid causes local oedema and together with active mucus secretion also induced by histamine results in the clinical signs such as rhinitis. The oedematous reaction is also apparent in the skin tests for allergens which is manifested as a fluid filled lump called a wheal (resulting from leakage of fluid from the blood stream) surrounded by a redness of the skin (erythema due to dilation of blood vessels), the so-called flare. Histamine also causes contraction of smooth muscle which can manifest itself as constriction of the bronchi resulting in difficulty in breathing or asphyxia.

There are some other mediators which are freshly synthesised when the mast cells are stimulated. These include leukotrienes LTC and LTD collectively called SRS-A (slow reacting substance of anaphylaxis) which has effects similar to histamine but are slower acting and more persistent and platelet activating factor (PAF) which can stimulate the formation of microthrombi in capillaries. This together with the over-reactive neutrophils result in tissue damage.

Π Before you read on, can you think of any ways in which you might test an individual for the presence of IgE specific for a particular allergen.

One simple way is by the simple prick test. A small amount of the allergen is inserted into the skin. Within a few minutes a wheal and flare reaction appears at the site of injection. You could also test the serum for specific IgE by the radio-allergosorbent test (RAST). Allergen attached to a cellulose disc is exposed to atopic serum and the amount of specific IgE bound is estimated by the binding of radio-labelled anti-IgE.

Systemic anaphylaxis

systemic anaphylaxis

Systemic anaphylaxis resulting from, say, an administered drug such as penicillin, induces constriction of the bronchi and accumulation of fluid in the lungs (mucosal oedema) leading to asphyxia. The vasodilation effects of histamine and other mediators results in circulatory shock due to lowered blood pressure and this is followed by collapse and finally death in an untreated patient. The drug epinephrine reverses the effects by being a vasoconstrictor and a bronchodilator and anti-histamines also effect recovery.

A visual summary of Type I anaphylaxis is given in Figure 7.12. Read through this figure careful as it contains a lot of information.

Basis for atopy

Why are some individuals sensitive to these allergens whereas the majority are not? The answer is we do not know. We do know that atopic individuals have high titres of serum allergen specific antibodies and of total serum IgE when compared to the normal population. It has been suggested that non-atopic individuals may produce high amounts of IgG allergen specific antibodies which successfully bind most of the allergen instead of the IgE on mast cells thus preventing the over-reaction observed in atopic individuals. Another possibility is that atopics have a deficiency in their suppressor mechanisms which would normally control the levels of IgE. Recent work has identified two suppressor molecules in mice, one of which down-modulates the low affinity IgE receptor (CD23) and the other suppresses IgE synthesis. There is also quite a lot of evidence to suggest that some forms of anaphylaxis have a genetic basis. For instance, a high frequency of HLA-B8 has been found in the atopic population.

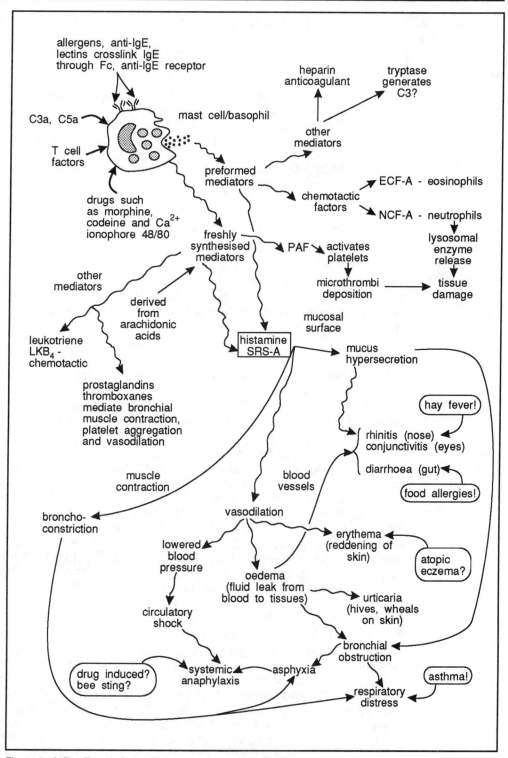

Figure 7.12 The Type I anaphylactic response and its effects.

Which of the following is/are correct?

1) Individuals can be sensitised by passive transfer of serum containing allergen specific IgE.

2) Degranulation of mast cells results in vasodilation, chemotaxis of neutrophils and platelet activation.

3) Erythematous and oedematous reactions can be induced by histamine.

4) Mast cells can be activated by anti-IgE, complement, morphine, anti-IgE receptor and allergens.

5) Systemic anaphylaxis results from the vasoconstrictive and bronchoconstrictive effects of histamine.

7.6.3 Type II - cytotoxic reactions

These reactions involve the binding of antibodies (IgG and IgM) to cell surfaces which results in complement activation and subsequent lysis of the cells or killing of the cells by effectors such as neutrophils possessing Fc receptors for IgG. Let us look at a few examples.

Transfusion reactions

This is a classic example of a Type II hypersensitivity reaction. Individuals of type O blood group possess IgM antibodies to both type A and type B red cells. If such an individual is transfused with type A blood, the IgM antibodies bind to the foreign red cells and activation of complement results in their destruction.

Haemolytic disease of the newborn

antibodies to Rhesus antigens

As well as the ABO blood group antigens, red cells of 85% of the population also carry Rhesus antigens. The individuals carrying the antigen are called Rh+, the remaining 15% with no Rhesus antigens are called Rh-. An Rh+ child born to a Rh- mother sensitises or primes the mother due to release of some red cells into the maternal circulation during birth. The mother produces antibodies, generally of the IgG class against the Rh+ red cells of the firstborn. In a future pregnancy, IgG antibodies will cross the placenta into the foetal circulation and bind to the Rh+ red cells. These opsonised red cells are then destroyed by foetal phagocytes and brain damage results from breakdown products of hemoglobin from the red cells.

Autoimmune haemolytic anaemias

cold agglutinin disease

Some individuals produce autoantibodies against their own red cells which results in destruction of the cells by complement mediated lysis or phagocytosis. This can lead to extreme loss of red cells and anaemia. Following some infections with mycoplasmas or during lymphoproliferative diseases, an IgM is produced against the individual's red cells which only binds to the cells at low body temperatures. The antibody is called a cold agglutinin as it agglutinates the red cells in the body extremities at cold outside temperatures and the disease is called a cold agglutinin disease. The coated red cells are lysed by complement in the warmer parts of the body. There are cold-reacting IgGs as well found in patients with chronic infections. A typical example is the Donath-Landsteiner antibody in patients with syphilis resulting in paroxysmal cold haemoglobinuria. As the name suggests, the patient has haemoglobin in the urine resulting in a dark urine due to lysis of the red cells.

Drug induced Type II reactions

drug induced cytotoxicity

Many drugs can act as haptens and bind to various cell types resulting in cell death. Some drugs, such as benzylpenicillin, undergo changes in the liver and this results in the generation of reactive groups which haptenise cells. Such reactions may lead to platelet loss (thrombocytopenia), leukopenia or agranulocytosis (loss of leukocytes) and, of course, anaemia.

Other autoantibodies in Type II hypersensitivity

autoantibodies

There are quite a few other autoantibodies which have been implicated in severe tissue damage. These include antibodies against the kidney glomerular basement membrane leading to kidney failure (Goodpastures syndrome), the acetyl choline receptor leading to severe muscle paralysis and loss of movement (Myasthenia gravis), pancreatic islet cells in some diabetics and the thyroid stimulating hormone receptor of thyroid cells leading to hyperthyroidism (Graves' disease).

7.6.4 Type III hypersensitivity - immune complex reactions

In the early days of passive immunisation, it was general practice for diphtheria patients to be treated with relatively large quantities of horse serum containing anti-diphtheria toxin antibodies. The patient produced antibodies to the proteins contained in the heterologous serum and immune complexes were formed in blood stream and other sites in the body. A repeat injection often resulted in death. The syndrome was called serum sickness. A similar condition can be induced by repeated drug administration resulting in production of IgG and/or IgM antibodies and complex formation.

∏ Before you read on, can you think of the mechanisms and possible damage caused by immune complexes in blood vessels?

Complexes, of course, are generated regularly during immune response and they are not necessarily harmful. Large complexes are removed by the liver while small complexes circulate for long periods without causing problems. Serum sickness is caused by the retention of immune complexes, often after a large or persistent antigen load, in minute blood vessels. This activates complement, which in turn induces basophils to release mediators such as histamine. This, as you know, induces increased vascular permeability and leads to deposition of the complex in the vessel wall. Neutrophils, attracted to the site (you should know how by now!) attempt to phagocytose the complex and, in so doing, spill out their lysosomal enzymes on to the vessel wall causing endothelial cell death and denudation of the basement membrane. Platelets attach to this denuded membrane causing microthrombi which can block the vessel. In kidney glomeruli this results in glomerulonephritis (as well as general tissue damage, this includes loss of proteins and red cells to the urine), in blood vessels

vasculitis

generally, it is called vasculitis.

∏ Construct a simple line diagram similar to Figure 7.12 on the mechanisms operative in Type III hypersensitivity. It will help you understand what is going on.

Similar mechanisms operate at other sites in the body. In rheumatoid arthritis, there is a lot of evidence for an autoantibody called rheumatoid factor which complexes with IgG in the joints causing damage to synovial membranes. In rheumatic fever, antibodies generated against group A *Streptococci* also bind to molecules bearing cross reactive epitopes in the heart muscle, cartilage and the kidney glomerular basement membrane. Workers exposed to mouldy hay generate antibodies to *actinomycete* spores and this results in immune complex deposition in the lungs with subsequent respiratory distress.

SAQ 7.11

Complete the following comparative table indicating your choice with a +.

Mechanism/Condition	Type I hypersensitivity	Type II hypersensitivity	Type III hypersensitivity
Complement activation			
Soluble immune complexes			
Mast cells or basophils			
Wheal and flare reaction			
Allergens			
Non-complement fixing Abs			
Histamine			
Neutrophil mediated damage			
Bronchoconstriction			
Involves IgG/IgM			
Red cell lysis			
Post-infection syndrome			
Leukopenia			
Drug induced			

Summary and objectives

You have now had a fairly detailed look at the anatomical sites where the cell interactions you have previously studied in depth in earlier chapters take place. You will remember our previous encounter with cell adhesion molecules which promote cell-cell interactions and we have now seen how these molecules have a further central role in promoting the movement of cells across anatomical barriers to sites of antigen deposition and inflammation. We have also looked at the effector cells of the immune response ie those cells that respond to signals derived from antigen specific T and B cells and migrate to the site of antigen and dispose of it. We then examined the kinetics of antibody production and how antibody mediated mechanisms can have deleterious effects on the host.

On the completion of this chapter you should be able to:

- describe the major structural features and roles of the lymphatic system and draw a fully labelled diagram of a lymph node;

- describe the cellular and molecular events in a lymph node during a humoral response and compare the structural components of spleen and lymph nodes;

- describe the main events in haemopoiesis and identify the cytokines responsible for selective development of haemopoietic cells;

- list the major characteristics and functions of neutrophils, basophils/mast cells and eosinophils;

- interpret experimental data on cell migration using homing receptors;

- identify the ligands for homing receptors on haemopoietic cells and describe the roles of cell adhesion molecules and cytokines in diapedesis;

- describe the kinetics of antibody production and distinguish between the properties of IgM and IgG produced during primary and secondary responses;

- explain the observed clinical symptoms observed during Type I hypersensitivity and compare the mechanisms involved in hypersensitivity reactions Types I - III.

Cell mediated immunity

Cell mediated immunity

8.1 Introduction

You have completed a detailed study of most aspects of humoral immunity and have learned that this provides, in the normal individual, good protection against extracellular antigens/pathogens. You will remember in Chapter 1 we referred to pathogens which were able to gain entrance to various types of cells in the host and thus were immune to the destructive effects of antibodies and complement. Alternative mechanisms have to be activated to deal with such pathogens. These mechanisms are collectively called cell mediated immunity (CMI) and are initiated and controlled by T cells. This is the subject of this chapter. It should be apparent that we cannot deal with this topic in the same detail as we did with antibodies and we are going to confine the discussion to some important aspects of CMI which will enable you to clearly distinguish between humoral and cell mediated mechanisms and to appreciate the contribution made by non antibody mediated mechanisms to the immune response. Additionally, we will demonstrate that, like antibodies, T cells can also mediate tissue damage. So, as well as providing protection against intracellular bacteria, viruses, fungi and parasites and dealing with host tumours, T cells are central players in many diseases such as autoimmune diseases, transplant rejection mechanisms and a host of hypersensitivity reactions collectively called contact hypersensitivity.

We shall begin by looking at ways for activating T cells and measuring T cell activity followed by the mechanisms of CMI. Finally we will take a brief look at some of the diseases which may be the result of T cell activity.

8.2 Activation of T cells

We have examined ways in which B cells are activated and their production of antibodies measured. Let us now look at some ways in which we activate T cells to assess their function.

8.2.1 T cell mitogens

PHA and Con A are T cell mitogens

There are various plant products called lectins which have an affinity for various cells including T cells. The two most commonly used lectins are called phytohemagglutinin (PHA) and concanavalin A (Con A). These are polyclonal activators as they activate the majority of T cells irrespective of their antigen specificity. They do this by binding to sugar residues on the TCR-CD3 complexes and this results in IL-2 production and proliferation of the T cells. This test is often performed in clinical laboratories to assess the functional competence of T cells.

assay for T cell activation

The laboratory test is performed as follows: a volunteer is bled into a heparinised tube (prevents blood clotting). The cells are then carefully layered on a cell separating medium called Ficoll-Isopaque and centrifuged. This results in a crude separation of blood cells into two fractions. The pellet consists of red cells and polymorphs and the cells at the interface of the blood and Ficoll consist of lymphocytes and monocytes.

This fraction is then washed free of Ficoll, resuspended in cell culture medium and exposed to mitogenic concentrations of the plant lectin for a period of 3 days. Radiolabelled thymidine is then added for a short period to the cell cultures. This will be taken up by actively dividing cells and incorporated into freshly synthesised DNA and the levels of uptake which directly correlate with the rate of cell proliferation, can be measured in a scintillation counter.

fractionation of cells from peripheral blood

There are ways of obtaining purer populations of T cells if you need them. Mixing T cells with sheep red cells (SRC) results in SRC binding to CD2 molecules on the T cell surfaces forming what we call rosettes. The rosetted T cells are much denser than non-rosetted T cells and are found in the pellet when the cells are subjected to Ficoll-Isopaque separation.

Monocytes/macrophages can be removed from cell populations by their adherence to plastic or by uptake of iron filings and subsequent removal using a magnet. You will also remember the panning techniques from Chapter 5. We should add that there are now other more sophisticated panning methods which we do not have time to examine.

Π Construct a flow diagram outlining a scheme for the purification of B cells by stepwise removal of firstly, polymorphs, then macrophages and finally T cells from a sample of human peripheral blood. Compare your scheme with that shown in Figure 8.1 which also gives the basic scheme for the T lymphocyte stimulation assay.

There are other reagents which also activate T cells in a similar fashion. For instance, antibodies against invariant or monomorphic regions of TCR-CD3 complexes promote mitogenesis but they may have to be fixed to inert carriers such as beads to effectively stimulate the T cells. Anti-CD2 antibodies also have a mitogenic effect on T cells.

superantigens

Recently another class of reagents have been identified as T cell stimulators. However, these can be distinguished from the true mitogens since they only selectively stimulate subpopulations of T cells which express certain V products. These have been called superantigens and they bind to MHC Class II molecules on the surfaces of APCs. They include staphylococcal enterotoxins which cause food poisoning and shock in humans and are highly mitogenic for T cells expressing $V\beta3$ and $V\beta8$ and streptococcal pyrogenic exotoxins. These superantigens stimulate T cells in a non-MHC restricted manner and do not need to be processed by APC.

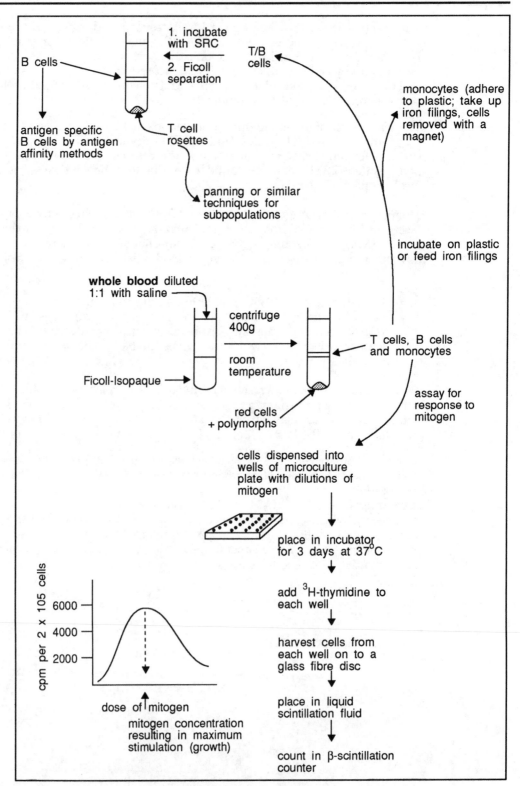

Figure 8.1 Separation of T cells and the T cell stimulation assay.

SAQ 8.1

Which of the following are correct/incorrect?

1) The T cell rosetting technique separated both CD4$^+$ and CD8$^+$ T cells from B cells on a Ficoll-Isopaque gradient.

2) Polymorphs are less dense than Ficoll-Isopaque.

3) T cells were treated with excess anti-CD2 antibodies and then rosetted with sheep red cells before being separated from B cells on a Ficoll-Isopaque gradient. The T cells, when placed in cell culture for a 3 day period were found to have proliferated.

4) Phytohemagglutinin or anti-CD3 antibodies activate both CD4$^+$ and CD8$^+$ cells.

5) Although superantigens do not need to be processed to activate T cells, they do require MHC II presentation.

8.2.2 Mixed lymphocyte reaction (MLR)

The mixed lymphocyte reaction is especially important in transplantation procedures when the clinicians wish to avoid allogeneic reactions between cells of the donor and recipient. It also has uses in assessing T cell competence as for mitogens. The basis for this test is that if you mix cells from two individuals, the T cells from each individual will respond to allogeneic MHC Class I and II molecules on the cells from the other individual as we saw in Chapter 1.

A relatively much higher proportion of your T cells, (sometimes more than 10% of the total cells), respond to allogeneic self MHC molecules on foreign cells than respond to MHC + peptide. This means that T cells with different antigenic specificities may be activated by the same allogeneic MHC I or II molecule. In other words, many different clones of T cells recognise allogeneic cells as foreign, because of their differences in the allogeneic MHC I or II molecules whereas only a few clones of T cells recognise a self MHC + peptide combination as foreign. This is probably because there are multiple different amino acid residues in the allogeneic MHC when compared to self MHC and these may appear similar to many different self MHC-peptide combinations recognised by individual specific T cell clones. Another way of thinking about this is that allogeneic MHC is 'more different' to self than a self MHC-peptide combination.

8.2.3 Laboratory procedure for mixed lymphocyte reaction

stimulators and responders

In the laboratory, it is normal procedure to inhibit the cells from one individual from responding by treating these cells with a drug such as mitomycin C or irradiation. The reaction is then called a one way MLR and the treated cells are called the stimulator cells and the untreated cells, the responder cells.

The cells are then mixed together in various proportions in tissue culture fluid and incubated at 37^{0}C for 4-6 days. Incorporation of radiolabelled thymidine added for the last few hours of the incubation is a measure of the level of cell proliferation due to allogeneic differences between the two population of cells. The proliferation results from recognition of allogeneic MHC II molecules by CD4 T cells resulting in expansion of these cells. If there are MHC Class I differences as well then CD8 cytotoxic cells will also fully expand.

8.2.4 Allogeneicity and responsiveness

However, as we discovered earlier, if there are only MHC Class I differences you would not expect any MLR as proliferation of CD8 cells requires IL-2 derived from activated CD4 cells.

Observations particularly from mice suggest it is not as clear cut as this. Proliferation and development of CD8$^+$ cytotoxic cells (CTLs) is strongest when there are differences in both MHC I and II. A strong proliferative response is also observed when there are just MHC II differences. When only MHC I differences are present there is a small proliferative response and CTLs do develop.

Π   Before you read on can you explain by some antigen processing mechanism how CTLs could develop in the absence of allogeneic differences in MHC II?

Unfortunately for you as the student, the division of CD4$^+$ and CD8$^+$ cells into IL-2 producers and cytolytic cells respectively is not absolute since some CD4 cells can be cytolytic and some CD8 cells can produce IL-2. There are two possible explanations for the above observations. Firstly, CD8 cells could produce low amounts of the required cytokines including IL-2 which would result in some proliferation and CTL development. Secondly, if allogeneic MHC I are being presented in professional APCs such as B cells, MHC I molecules which are endocytosed from the surface can be processed and reappear as self peptides associated with MHC II molecules thus stimulating CD4$^+$ cells.

Π   This second explanation represents a much less efficient process than if allogeneic MHC II molecules were expressed. Can you think why?

Did you work it out? If a T cell is exposed to a cell expressing allogeneic MHC II, then all of the MHC II molecules are capable of stimulating the T cells. If the cell is expressing syngeneic MHC II then only some of them, will be presenting processed MHC I peptides which capable of stimulating the T cells and a lower response will ensue.

8.2.5 Cell mediated lympholysis assay

assay for CD8$^+$ cells, cell mediated lympholysis

The expansion of CD8 cells can be assessed by an assay which can be performed alongside the MLR. It is called the cell mediated lympholysis (CML) assay. The 6 day cells are exposed to target cells (expressing allogeneic MHC I) which have been incubated in a solution of radiolabelled chromate (^{51}Cr). This is taken up into the cytoplasm of the target cells. When the cells are lysed by the CD8$^+$ cells, the amount of radiolabel released can be measured and the percentage target cell lysis assessed.

In the clinical laboratory, previously stimulated responder cells are used in what is known as the primed lymphocyte test (PLT). The response in this case is much stronger and more rapid and the test is complete in 3 days instead of 6 days.

We have illustrated the phenomena of MLR/CML in Figure 8.2 which also shows the basis of the two laboratory assays.

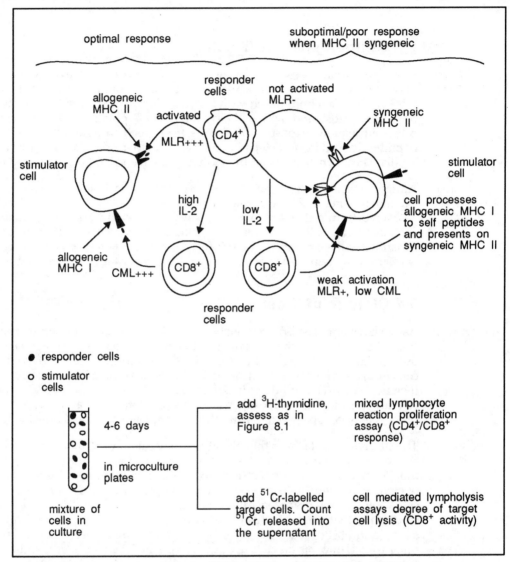

Figure 8.2 The mixed lymphocyte reaction (MLR) and cell mediated lympholysis (CML) assays.

SAQ 8.2

Match items in the left with those in the right column using each item only once. + indicates a low response, ++ indicated an optimal or strong response, - indicates no response. MLR in your response indicates CD4 proliferation only, CML indicates CD8 proliferation and development of CTL activity.

1) syngeneic MHC I and MHC II   A) MLR++, CML++

2) allogeneic MHC I, syngeneic MHC II B) MLR++, CML-

3) allogeneic MHC I, allogeneic MHC II C) MLR+, CML+

4) syngeneic MHC I, allogeneic MHC II D) MLR-, CML-

8.3 Antigen specific T cells

primed T cells
from *in vivo*
immunisation

Most individuals possess memory T cells derived from previous encounters with a variety of antigens or as a result of deliberate immunisation. For instance, hospital workers are often immunised with a *Mycobacterium tuberculosis* vaccine to protect them from tuberculosis and these will have increased population of *M. tuberculosis* specific T cells. Subsequent exposure of T cells from such individuals to APCs expressing peptides derived from such pathogens will lead to re-stimulation of the antigen specific T cells. These can be assessed by ^{3}H-thymidine incorporation or by IL-2 release.

Antigen specific T cells can also be recovered from lesions of patients and clonally expanded using IL-2. For example, tumour specific CD8$^+$ cytotoxic T cells have been recovered from cell suspensions prepared from solid tumours by incubation in IL-2 containing cell culture medium. By continued exposure to IL-2, these cells can be expanded to huge numbers and reinfused into the patient from whom the solid tumour was surgically removed in an effort to kill any metastatic growths in the patient.

8.4 Cloning of T cells

As we have emphasised previously, T cell cloning was facilitated by the discovery of IL-2. The basis of cloning is relatively straightforward although each laboratory has its own variation of the general methodology. It is the usual practice to obtain primed T cells from individuals or experimental animals as we have just indicated. These cells are then exposed in cell culture to irradiated APCs presenting the peptides for which the T cells are specific. This will result in selective proliferation of antigen specific T cells. This process may be repeated to ensure that only antigen specific T cell clones survive.

The surviving T cells from the cell culture may contain a number of different clones and the next procedure involves separating the individual clones thus ensuring monoclonality. The procedure is called cloning by limiting dilution. The mathematics of cloning and the procedures are rather complex and we will simplify it as follows. A cell suspension is prepared from the mixture of T cells and is diluted in cell culture medium so that when it is dispensed into 0.2ml wells of a 96 well cell culture plate, the average number of cells in each well is 0.5 cells. In other words, there is one cell in every other well. Using this procedure there is a very high statistical probability that the T cell subsequently recovered from a single well represents a single clone. The wells also contain some irradiated feeder cells (spleen cells or thymus cells) which supply essential cell contact and various nutrients to the single cell to promote its growth. IL-2 is also present to support expansion of the clone.

These T cell clones can be deep frozen in liquid nitrogen and stored for many years. We shall see later how such T cell clones provide information on T cell involvement in disease.

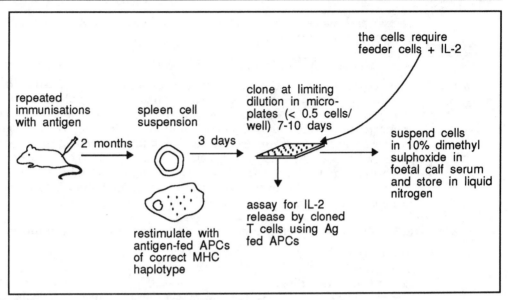

Figure 8.3 T cell cloning using murine cells.

SAQ 8.3

Which one of the following statements is incorrect?

1) During an MLR there are many different CD4$^+$ and CD8$^+$ T cells responding to the target cell.

2) The mixed lymphocyte reaction does not require previous T cell priming.

3) The low CML response to cells allogeneic for MHC I and syngeneic for MHC II may be due to lower levels of IL-2 released by T cells activated by MHC II-self MHC I peptide compared to activation by allogeneic MHC II.

4) Tumour specific T cells removed from a primary tumour proliferate in response to IL-2 in the absence of tumour antigen.

5) Mixing of a patient's tumour cells and T cells results in proliferation similar to a mixed lymphocyte reaction *in vitro*.

8.5 The roles of T cells in antibacterial immunity

We will define antibacterial immunity as immunity to intracellular bacteria, fungi and parasites. Let us examine what roles T cells play in protective immunity to these pathogens.

This type of immunity is involved in response to a wide variety of intracellular pathogens. These include bacterial species *Mycobacterium tuberculosis*, *Salmonella typhosa*, *Brucella abortus* and *Listeria monocytogenes*, some fungi including some *Candida* and *Histoplasma* species and protozoa including *Toxoplasma*, *Leishmania* and *Trypanosoma* species and malarial plasmodia. Notice, all these are intracellular and therefore fully protected from antibodies.

8.5.1 The adoptive transfer of immunity to *Listeria monocytogenes*

The original description of cell mediated immunity was derived from the following experiment. Mice were immunised with a small dose of virulent *Listeria monocytogenes*. A few days later, the animals were sacrificed and either serum or spleen cells were transferred to nonimmune recipients. These were subsequently challenged with a lethal dose of *Listeria* and/or tubercle bacillus. The animals given immune serum died as did those animals challenged with tubercle bacillus alone. The animals given immune splenocytes and subsequently challenged with either *Listeria* alone or both bacterial species survived. This experiment demonstrated that antibodies were not protective but immune splenocytes were and the immunity was cell mediated.

8.5.2 Possible cellular mechanisms in antibacterial immunity

The *Listeria* story

Let us take another look at the *Listeria* story (Figure 8.4). The *Listeria* injected into the donor animals would result in priming of *Listeria* specific T cells. On transfer to recipients and subsequent challenge with *Listeria*, infected macrophages would process and present some *Listeria* components along with MHC Class II. This combination would be seen by the memory T cells and the latter would activate the infected macrophages resulting in the killing of the intracellular bacteria. You will notice, however, that tubercle bacillus injected along with the *Listeria* were also destroyed even though there were no memory T cells specific for this bacteria. This apparently nonspecific phenomenon was due to the ability of the activated macrophages to kill the intracellular bacteria irrespective of the species due to the stimulation by *Listeria* specific CD4$^+$ T memory cells.

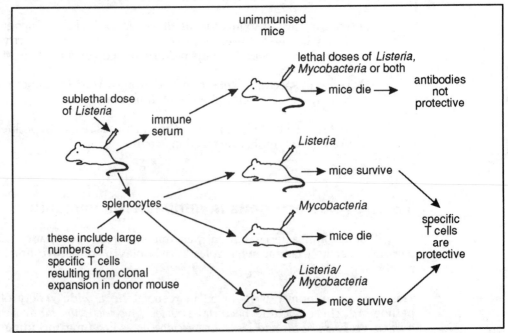

Figure 8.4 Specific activation of CD4$^+$ T cells leads to nonspecific protection *in vivo*.

8.5.3 The classical CD4⁺ T cell mediated mechanism

We have to warn you that our ideas on antibacterial (including all other intracellular non-viral pathogens) immunity is in a state of some confusion at the moment firstly because we are trying to decide which cytokines are involved and secondly because we now think that both CD4⁺ and CD8⁺ cells are involved.

basic CD4⁺
mediated
mechanism,
cytokine
production

The updated classical mechanism is illustrated in Figure 8.5. The macrophage phagocytoses the bacteria and some of these are endosomally processed and expressed along with MHC II on the cell surface. Primed CD4⁺ T cells with specificity for MHC II + bacterial peptide crosslinks by its TCR-CD3 complexes (remember the other cell adhesion molecules as well!) Activation then follows the same sort of path shown in Figure 6.8 with the macrophage production of IL-1 which activates the T cell but also acts on local endothelial cells along with macrophage derived TNF to increase expression of ICAM-1, VCAM-1 and ELAM-1 and induce production of cytokines such as IL-8. The T cell produced IFN-γ which enhances expression of MHC II on the macrophage thereby increasing presentation but which also promotes increases in the bactericidal activities of the infected cell. This may result in killing of the bacteria.

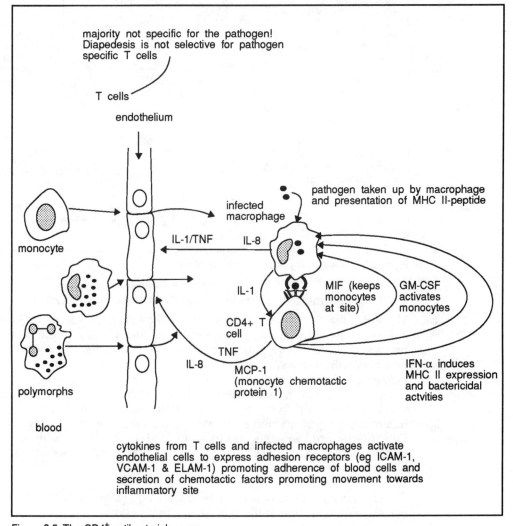

Figure 8.5 The CD4⁺ antibacterial response.

IL-8 and MCP-1   The inflammatory response is enhanced by production of chemotactic factors such as
                  IL-8 and MCP-1 (monocyte chemotactic protein 1) which attracts more monocytes to the
                  site as well as T cells and neutrophils. The T cell also produces macrophage migration
                  inhibition factor (MIF) which prevents macrophages leaving the site and possible
                  GM-CSF which is an activator of both macrophages and polymorphs.

                  Since the macrophages are such important cells in CMI and indeed also in humoral
                  immunity we have summarised the major features of these cells in Figure 8.6.

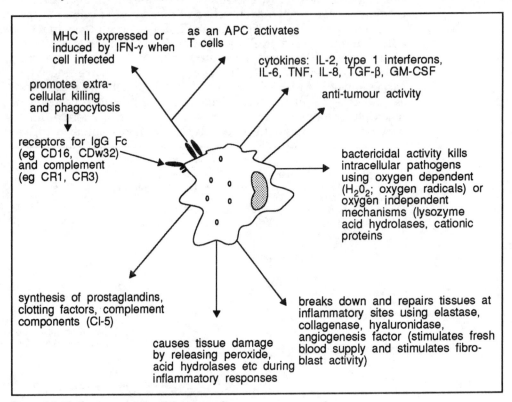

Figure 8.6 The major properties of the macrophage.

8.5.4 The delayed type hypersensitivity reaction (DTH)

DTH skin   The sort of reaction described above can be elicited by intradermal injection of a suitable
reaction   antigen or by skin painting. A typical reaction in many is the so called tuberculin
                  reaction. Persons working in the medical professions need protection against possible
tuberculin   infection with *Mycobacterium tuberculosis* so they are examined for the presence of
reaction   primed T cells with reactivity to this bacteria. A partly purified extract of these bacteria
                  called purified protein derivative (PPD) is injected into the skin. If the person has been
                  previously exposed to the bacteria ie has recovered from or been vaccinated against
induration and   tuberculosis a skin reaction is observed, a reddening of the skin (erythema) and a hard
eythema   lump (induration) which is maximal about 2 days after the injection. This delayed type
                  hypersensitivity reaction (DTH) contrasts with the hypersensitivity reactions types I-III
                  which are more immediate.

                  The erythema is due to dilation of blood vessels and the induration is due to the influx
                  of cells and deposition of fibrin at the site of injection.

SAQ 8.4

Which one of the following items is not necessary for the DTH response?

1) CD4$^+$ T cells.

2) Endosome.

3) IFN-γ.

4) ICAM-1.

5) IL-2.

8.5.5 Is CD4$^+$ the major cell mediating antibacterial immunity?

cytolytic CD4$^+$
T cells

In mycobacterial infections in Man, there is little evidence for killing of the bacteria and experimental evidence supports the idea that CD4$^+$ cytolytic cells actually lyse the infected cells in a MHC II restricted manner. You can imagine that this could be advantageous to the infected host since the T cells are destroying the habitat of the bacteria. It may be especially useful for cells such as hepatocytes which become infected by *L. monocytogenes* and Schwaan cells infected with *Mycobacterium leprae*. These cells have poor antimicrobial activities and cannot dispose of the bacteria. Lysis of the cells could result in release of the bacteria which could then be taken up by professional phagocytes attracted to the site by cytokines and killed. Antibodies are produced in mycobacterial infections and could possible aid in the destruction of some of these released bacteria although there is little evidence for this.

SAQ 8.5

Which of the following are common to both innate immunity and DTH reactions?

1) Diapedesis.

2) Chemotactic factors.

3) T cells.

4) Antigen specificity.

5) Macrophages.

8.5.6 CD8$^+$ T cells and antibacterial immunity

Intracellular pathogens primarily reside in the endosomal compartment of host cells which would inevitably lead to processing and presentation on MHC II molecules thus activating CD4$^+$ T cells. This fits the classical dogma of CD4$^+$ T cells being the major mediators of antimicrobial immunity. However, there are reports that suggest that the CD4$^+$ T contribution to protection against such pathogens as *Listeria* may be minor and there is good evidence to suggest that CD8$^+$ T cells may also be implicated.

Π   It would be a good idea to revise the mechanisms of antigen processing for both endogenous and exogenous antigens (Sections 4.5.7 and 4.5.8 before you read on).

CD8⁺ T cells
may lyse
bacterially
infected cells

Since CD8⁺ T cells recognise peptides bound to MHC I molecules and that these peptides are produced in the cytoplasm by proteasome activity before being transported to the endoplasmic reticulum for attachment to MHC I molecules, the bacteria must have been able to exit from the endosomes into the cytoplasm to be subsequently processed. A mechanism has been suggested for *Listeria monocytogenes*. These bacteria produce listeriolysin which perforate the endosomal membrane facilitating escape of the bacteria into the cytoplasm thus evading the bactericidal effects of the endosomal enzymes. This action then provides the source of proteins for MHC I presentation and subsequent activation of CD8⁺ T cells resulting in lysis of the infected cells (Figure 8.7).

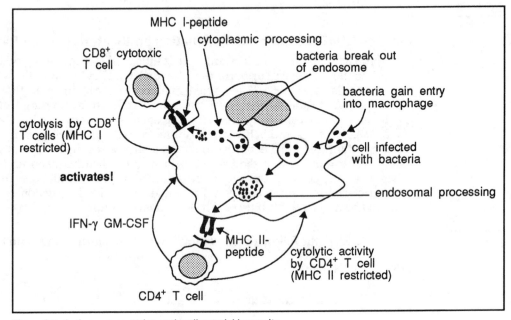

Figure 8.7 Antigen presentation and antibacterial immunity.

The suggested lytic activities of both CD4⁺ and CD8⁺ T cells may be beneficial to the host by depriving the bacteria of a protecting environment with subsequent killing perhaps by cells with better bactericidal properties. Some of these bacteria do have difficulty surviving outside these cells. It would be expected, of course, that lytic activities can also result in tissue damage and possible dissemination of the bacteria to other host sites.

∏ From Figure 8.7, write down three mechanisms the infected cell could use to dispose of the pathogen.

In summary, there are, then, serious suggestions that the classical concept of CD4⁺ T cells activating cells infected with intracellular non-viral pathogens and thus causing the demise of the pathogens may not hold in some infections. Evidence suggests that both CD4⁺ and CD8⁺ T cells are involved and these activities would result in the destruction of the infected cells in a similar manner to virally infected cells.

From Figure 8.7 you should have been able to conclude that 1) the infected cell could be activated by CD4 T cells specific for bacterial peptide-MHC II to kill the pathogen using lysosomal enzymes - in this case the infected cell survives; 2) the CD4 T cell may kill the infected cell by MHC II restricted cytolysis or 3) escape of the bacteria from the endosome promotes cytoplasmic processing and activation of CD8 CTL which kills the infected cell.

8.5.7 Involvement of NK cells and γδ cells

Additionally, there are suggestions that Natural Killer cells and TCR-1 bearing cells are also capable of lysing these infected targets. There are extraordinarily high numbers of γδ-T cells which can be activated by mycobacterial components such as heat shock protein 60 and it is interesting to note that these T cells are sited for the most part in the submucosa where a majority of pathogens gain entrance.

SAQ 8.6

Which of the following statements are correct?

1) Listeriolysin promotes CD8$^+$ T cell activity.

2) The DTH mechanism results in destruction of intracellular pathogens.

3) MHC I and II presentation can result in lysis of bacterially infected macrophages by the same CTL.

4) CD4$^+$ T cells may be clonally expand when mediating the DTH reaction.

5) IL-1 and TNF produced solely by infected macrophages activate the local endothelial cells to express high numbers of adhesion molecules.

8.6 Cell mediated immunity to viruses

Most responses to viruses have both a humoral and a cell mediated immune component. Antibodies could be important in immunised individuals in preventing the entry of the viruses into cells and may enhance phagocytic clearance of viruses. We are going to briefly examine the natural immune cellular mechanisms to viral infection and then the virus specific response by cytotoxic T cells.

8.6.1 Natural immunity of viral infection

type I
interferons

Many cells infected with a virus synthesise cytokines called Type 1 interferons which consist of two groups of polypeptides. One group, IFN-α, consists of a family of about 20 closely related proteins of molecular weight 18kD. The principal source of IFN-α is the macrophage and this group has been called leukocyte interferon to denote the source. The other group has only one member, IFN-β produced in the laboratory by fibroblasts and hence called fibroblast interferon. However, as we said earlier, many cells produce Type 1 interferons on being infected with virus and the secreted products bind to a common receptor on neighbouring cells. These interferons promote the synthesis of a number of antiviral enzymes which then prevent the spread of the virus from the infected cell.

∏ Can you remember the term for this type of cytokine activity?

Since these interferons mediate their action on neighbouring cells, it is termed paracrine activity.

Type 1 interferons also enhance the expression of MHC I molecules on infected cells and activate natural killer cells to lyse the infected cells. So you can see that these activities promote the eradication of the virus.

8.7 Cytotoxic T cells in viral infections

8.7.1 CD8$^+$ differentiation to an effector CTL

CD8$^+$ cytotoxic T cells, after maturing in the thymus, require like T helper cells, a priming mechanism in order to differentiate to a fully functional CTL in the periphery. As we mentioned earlier, the virus specific pre-CTL binds to MHC Class I-peptide on a virally infected cell. In order to differentiate into a mature CD8$^+$ CTL it requires IL-2 and possibly other cytokines such as IFN-γ and IL-6 which it may not make in sufficient quantity. The source of these mediators will be CD4$^+$ T cells which will need to be bound and activated by MHC II-peptide on the same cell! In many cases, the infected cell will not be a professional APC ie it will not be expressing MHC II and will need stimulating with IFN-γ. It would seem logical that the CD8$^+$ T cell must provide this initial stimulus since CD4$^+$ T cells have to first be activated by MHC II-peptide (Figure 8.8). This is however speculative.

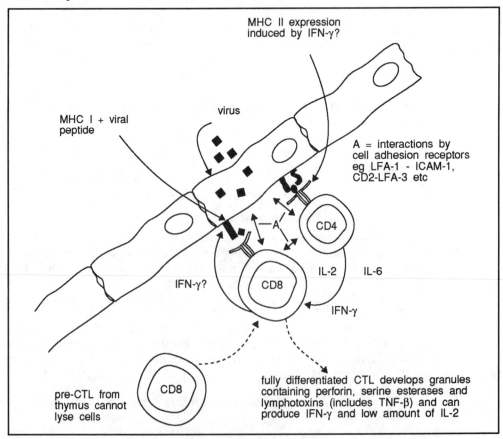

Figure 8.8 Differentiation of CTL (see Sections 8.7.1 - 8.7.4 for details).

Actually, this classical dogma of a requirement for CD4$^+$ cells to aid in priming of CD8$^+$ T cells may not be a general requirement. For example, it would appear that there is little requirement for CD4$^+$ T cells for the priming/activation of CD8$^+$ T cells in influenza virus infections.

SAQ 8.7

CD8$^+$ T cells require which of the following for differentiation to a fully functional CTL?

1) A classical antigen presenting cell.

2) Exogenous antigen.

3) Type 1 interferons.

4) Lysis of the target cell.

5) Exogenously derived cytokines.

8.7.2 CD4$^+$ and CD8$^+$ CTL

There is little doubt that the major CTL is of the CD8$^+$ phenotype. However, the immune system appears to be flexible in some viral infections in its use of CD4$^+$ CTL if there is either a lack of CD8$^+$ T cells or a deficiency of MHC I binding to available viral peptides thus resulting in no presentation. For example, CD8$^+$ CTL clearly provide the protective immunity against murine cytomegalovirus in normal mice with no evidence of CD4$^+$ CTL. If mice are depleted of the CD8$^+$ T cell effectors, however, CD4$^+$ T cells assume the responsibility of protective immunity. There is also evidence for CD4$^+$ CTL when viral peptides can only be presented on MHC II molecules.

8.7.3 Viral entry and presentation

With regard to phenotype and effector function, it is necessary to examine the ways in which viruses enter cells because this gives us clues as to which pathway is being utilised leading to either MHC I or II presentation. Enveloped viruses probably enter cells either by fusion with the cell membrane or receptor mediated endocytosis. Non-enveloped viruses use exclusively the latter pathway. Hence viruses may multiply in the cytoplasm producing viral proteins for MHC I presentation and CD8$^+$ activation or may be confined to the endosomal compartment which will lead to MHC II presentation and CD4$^+$ activation. Of course, CD4$^+$ activation will usually result in provision of help to a neighbouring CD8$^+$ cell; the actual activation of CD4$^+$ cells to become cytolytic by MHC II restricted mechanisms is probably fairly rare except in the sort of circumstances referred to above. We should also remember that free virus or parts of the virus discarded on entry into the cell could also be processed via the MHC II pathway.

8.7.4 Cell mediated lysis

the lytic mechanism

Once primed with the help of CD4$^+$ T cells, CD8$^+$ CTL develop granules containing perforin or cytolysin (related to complement C9) and degradative enzymes and develop the capacity to produce cytokines such as IFN-γ, lymphotoxin (TNF-β) and low amounts of IL-2.

perforin

The lytic mechanism is initiated by binding of the CTL to the target cell. In much the same way as we described cell interactions, this involves LFA-1, CD2 and CD8 on the

CTL binding to ICAM-1, LFA-3 and MHC I respectively on the target cell together with TCR-CD3 crosslinking with MHC I viral peptide. The granules then release the pore-forming protein (perforin) which attaches to the target cell membrane and polymerises in the presence of calcium to form transmembrane channels. The CTL then releases the degradative enzymes (serine esterases) and lymphotoxins which enter the cell through the channels to cause irreversible damage including DNA fragmentation (apoptosis) induced by TNF. IFN-γ released from the CTL is also probably involved in anti-viral activities in the target cell as indicated for the Type 1 interferons.

Thus the CTL destroys the habitat for replication of the virus and its products may destroy viral particles. Alternatively, viruses released from the dying cells could be cleared by antibodies. Figure 8.9 illustrates the events in cell mediated lysis by CTL.

SAQ 8.8

Which one of the following statements is correct?

1) All CD8$^+$ T cells produce IL-2.

2) CTL mediated lysis is solely the result of MHC I recognition.

3) Binding of cell adhesion molecules precedes lysis by CTL.

4) CTLs cannot attack CD4$^+$ T cells.

5) Complement and CTL mediated killing are very similar mechanisms.

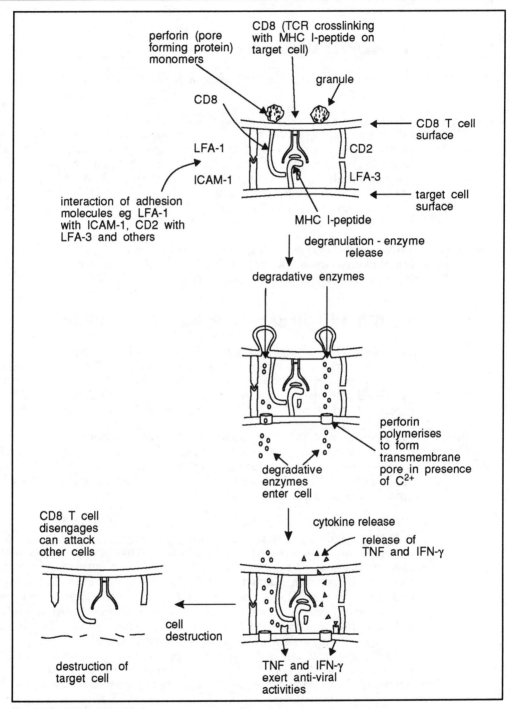

Figure 8.9 CTL mediated lysis.

8.8 Natural killer cells

∏ Remind yourself of the basic characteristics of NK cells in Chapter 7 before you read on.

NK cells and virally infected cells

As we said earlier, NK cells are capable of lysing cells infected with some viruses as well as some tumour cells. NK cells have a similar lytic mechanism to CD8⁺ T cells producing pore forming proteins, cytotoxins and enzymes for destruction of the target. They are, however, not thought to produce lymphotoxin. NK cells are the principle cells which mediate an alternative form of killing called antibody dependent cellular cytoxicity (ADCC). This involves the binding of its low affinity IgG receptor CD16 to IgG molecules on target cells followed by the lytic mechanism involving perforin already described.

lymphokine activated killer cells

When stimulated with high concentrations of IL-2, NK cells differentiate into lymphokine activated killer cells (LAK) which exhibit increased cytolytic activity and kill a much broader range of tumour targets than NK cells.

8.9 CTL and other cells in tumour immunity

Cells able to kill tumours must be able to recognise tumour cells which suggests that the latter must express molecules not recognised as self ie tumour antigens. These can take various forms including virally derived antigens and oncofoetal antigens (proteins expressed during development only). Whatever form they do take, they need to be processed and presented on MHC I or II molecules to stimulate T cells.

tumour infiltrating lymphocytes

Cells expressing tumour specificity can be recovered from cell suspensions of primary tumours from cancer patients and these cells also exhibit MHC restriction. They are referred to as tumour infiltrating lymphocytes (TIL). Such cells kill in the same way as virus specific CTL in the laboratory and CTL have been shown to be effective against virally induced tumours in mice.

NK and tumour immunity

Although you do not find may NK cells in these tumour infiltrates, it is interesting to note that T cell deficient mice with normal levels of NK cells do not exhibit a higher incidence of tumours suggesting that these cells may mediate some sort of protection against spontaneous tumours. In the laboratory, NK tumouricidal activity is greatly increased by cytokines such as IL-2, TNF, Type 1 interferons and immune interferon and it may be that NK cells may be active *in vivo* after stimulation by such cytokines provided by T cells and macrophages. You have already heard that NK cells exposed to IL-2 become much more tumouricidal as LAK cells. Immunotherapy of patients with advanced cancers by infusion of huge numbers of LAK cells + IL-2 have had some success resulting in regression of some types of tumour but the latest approach is to use tumour specific CTL and these trials are proceeding at the moment. We do not know what it is about tumour cells which make them targets for NK and LAK cells.

macrophages and tumours

Macrophages have the right armoury to be effective cells in destruction of tumours especially those coated in anti-tumour antibodies. These cells are also a major source of TNF-α and TNF-β (lymphotoxin) both of which appear to have the ability to destroy tumour cells. Thus, macrophages could kill tumour cells by phagocytosis and by an extracellular mechanism. Tumour specific CD4⁺ T cells may be involved in directing macrophages to the tumour; recent evidence also suggests that CD4$^+$ T cells may destroy tumours by secretion of cytotoxins.

We do not know very much about tumour immunity *in vivo* in Man although the pace of discovery has quickened considerably recently with the identification of tumour specific MHC restricted CTL in primary tumours and the many cytokines which will probably be useful in immunotherapy of the future. The main purpose of this section is simply to show that there is a considerable potential cell mediated response to tumours and to wet your appetite with this very exciting area of research.

| SAQ 8.9 | Indicate by a + in the table which cells possess any of the mechanisms shown in the left column which could contribute to anti-tumour defence. |

Anti-tumour mechanism	NK	CTL	Macrophage
1. Phagocytosis			
2. Perforin mediated lysis			
3. Lymphotoxin			
4. Tumour specific mechanisms			
5. Antibody mediated killing			
6. Extracellular killing			
7. Complement mediated killing			

8.10 Diseases caused by T cells

We are sure you have been impressed by the importance of the CD4$^+$ T cell in both humoral and cell mediated immunity; indeed it has been called the central cell in immunity. However, just like antibodies which can also mediate tissue damage, T cells also cause hypersensitivity reactions. In addition, T cells may become autoreactive resulting in autoimmune diseases or may respond to chronic antigen stimulation and cause severe tissue injury. Finally, T cells appear to be responsible for the rejection of transplants (allografts).

For the remainder of this chapter, we are going to introduce you to the mechanisms behind:

- contact hypersensitivity;

- granulomatous diseases;

- autoimmune diseases;

- transplantation reactions.

8.10.1 Contact hypersensitivity (delayed type hypersensitivity-Type IV)

contact allergens are haptens,the priming reaction

Some forms of dermatitis are non-immunological being caused by general skin irritants whilst other forms are due to IgE mediated responses which we have dealt with earlier. The third form is due to small molecular weight lipid soluble antigens (haptens) which are able to penetrate the epidermis to the deeper layers where they bind to tissue proteins or cells. The haptenised proteins will be processed in cells such as Langerhans cells, carried to the local lymph node paracortex to become IDC and will prime specific CD4$^+$ T cells resulting in clonal expansion. These sensitised T cells will, if exposed to the haptenised peptide again, elicit the contact hypersensitivity reaction.

Most individuals can be sensitised to at least some of these environmental haptens and the eliciting dose can be far less than the sensitising dose. We have indicated the nature of some of these contact sensitising agents in Figure 8.10.

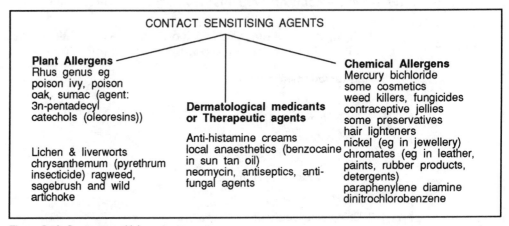

CONTACT SENSITISING AGENTS

Plant Allergens
Rhus genus eg poison ivy, poison oak, sumac (agent: 3n-pentadecyl catechols (oleoresins))

Lichen & liverworts chrysanthemum (pyrethrum insecticide) ragweed, sagebrush and wild artichoke

Dermatological medicants or Therapeutic agents

Anti-histamine creams local anaesthetics (benzocaine in sun tan oil) neomycin, antiseptics, anti-fungal agents

Chemical Allergens
Mercury bichloride some cosmetics weed killers, fungicides contraceptive jellies some preservatives hair lighteners nickel (eg in jewellery) chromates (eg in leather, paints, rubber products, detergents) paraphenylene diamine dinitrochlorobenzene

Figure 8.10 Contact sensitising agents.

the elicitation reaction

The elicitation reaction proceeds as follows. Hapten conjugation probably occurs within the first 4 hours and sensitised T cells start to arrive within 8 hours and become activated by the hapten conjugated to presented peptide on cells in the dermis, probably macrophages or Langerhans cells. Cytokines released by both the T cell and the macrophages cause an influx of neutrophils followed by monocytes (see Figure 8.5 for the mechanism). TNF, released by both cell types is thought to induce some plasma leakage which results in some oedema at the site. This may also be caused by serotonin released by mast cells which are activated by putative T cell factors and would cause erythema of the skin. The neutrophils and monocytes cause local damage to tissues leading to vesicles formation about 36 hours after antigen contact. These vesicles may rupture leading to spreading of the lesion. Regeneration of the site by macrophages and fibroblasts is usually complete 4 days into the reaction.

induration and erythema

The controlled DTH skin test resulting in an induration and erythema is usually maximal 2 days after application of the antigen. In some hypersensitive individuals, the skin test also results in vesicle formation.

SAQ 8.10

Complete the comparative table of Type I and IV hypersensitivity reactions.

	Type I	Type IV
1) Named reaction		
2) Mediated by		
3) Time of reaction		
4) Skin test result		
5) Major histology		
6) Passive transfer		
7) Effector molecules		

8.10.2 Granulomatous reactions

chronic DTH reaction, granulomas, epithelioid and giant cells

When antigen persists at an inflammatory site, CD4$^+$ T cells continue to be activated and release cytokines. This leads to what we may call a chronic delayed hypersensitivity reaction and tissue damage. Continued stimulation of the macrophages induces changes in these cells such as increased cytoplasm and organelles until they look like epithelial cells in the skin and they have been called epithelioid cells. These may also fuse to become multinucleate giant cells. It is thought that the macrophages also produce platelet-derived growth factor which stimulates proliferation of fibroblasts leading to fibrous tissue and also induces collagen synthesis. There is also much cell damage by macrophages and these cells themselves may be killed by antigen specific T cells (CD8$^+$ and CD4$^+$) within the inflammatory area resulting in necrosis. The resulting granuloma consists then of a centre of necrosis surrounded by epithelioid and giant cells and fibrous tissue and possibly represents an attempt by the host to wall off the infectious agents it cannot get rid of.

The types of organisms that promote formation of granulomas include *Mycobacterium tuberculosis* and *M. leprae*, *Leishmania* and *Listeria monocytogenes*. The extensive tissue damage that occurs in response to pathogens such as *M. tuberculosis* in untreated patients leads to partial loss of organ function such as the lungs in pulmonary fibrosis.

8.10.3 Autoimmune disease

There are a host of diseases in which the mechanisms of self recognition have broken down resulting in activation of CD4$^+$ T cells and production of autoantibodies or autoreactive cells. There are many theories as to how autoimmunity comes about but we are not going to deal with those. We are simply going to introduce you to an experimental model which strongly implicate T cells as the culprits in development of autoimmune disease and which will show you the approaches used by researchers to elucidate the mechanisms involved and how they lead to ideas on immunotherapy.

Experimental allergic encephalomyelitis

myelin basic protein, lymphocyte and macrophage infiltration into nervous tissue

This is an experimental model of an organ specific autoimmune disease mediated in large part by CD4$^+$ T cells. The white matter of the nervous tissue consists largely of nerve fibres covered in a myelin sheath of which the principal constituent is myelin basic protein. You can induce experimental allergic encephalomyelitis (EAE) in mice by injecting homogenised brain tissue or myelin basic protein (MBP) in an adjuvant (this promotes uptake by APCs and macrophages). About two weeks later the animals develop hind leg paralysis and the neurologic lesions show lymphocyte and

macrophage infiltration into nervous tissue accompanied by demyelination of the
nervous tissue including the brain.

SAQ 8.11

Which of the following would represent evidence for the involvement of T cells
in a demyelinating disease?

1) Presence of T cells at the site of damage.

2) T cells from the lesion proliferate when presented with myelin basic protein.

3) Anti-CD4 antibodies prevent onset of disease.

4) Cloning of T cells from the lesion shows predominance of T cells with
   specificity for a proteolytic fragment of myelin basic protein.

5) T cells from the lesion proliferate in response to mitogenic stimulation.

anti-T antibodies implicate CD4⁺ T cells

Let us summarise the evidence which implicates $CD4^+$ T cells as the primary disease
inducing component of EAE. Anti-CD4 antibodies were found to prevent EAE and EAE
symptoms could be induced in non-immunised mice by transfer of MBP-specific T cells.
It was subsequently found that the majority of MBP-specific T cells expressed Vβ8 gene
segment, the remainder Vβ13 together with only 2 Vα segments suggesting the
induction of EAE was due to a very restricted number of $CD4^+$ T clones. Anti-idiotypic
antibodies, prepared against Vβ8 and Vβ13 expressing T cells, were found to inhibit
MBP-specific lymph node proliferation responses. When injected into mice, the
antibodies significantly diminished the response of MBP and actually prevented the
onset of EAE symptoms when administered prior to immunisation with MBP.

Π We think you will agree that the evidence for $CD4^+$ T cell involvement is quite
strong. Can you think of the possible mechanism by which the antibodies mediate
their effects (think of antigen presentation)?

TCR peptides from autoimmune T cells protect against EAE

Since there were so few Vβ segments expressed in encephalitogenic T cells, it was
possible to construct peptides representing the CDR II or CDR III sequences of some of
these Vβ or Vα chains. Immunisation of rats with these peptides resulted in antibodies
and DTH skin reactions and protected the rats from EAE induction. The antibodies
would protect by binding to the TCR of the $CD4^+$ T cells thus blocking TCR interactions
with the MBP⁻ peptides being presented by MHC II on APC's. Additionally, MHC I
restricted T cells specific for the Vβ8 peptide could be isolated from lymph nodes of
protected rats which could confer protection to other animals. It was also noticed that
T cells could be isolated and expanded from the lymph nodes of protected animals
which when adoptively transferred to naive rats induced the disease.

You may well say that EAE is not representative of an autoimmune disease because the
self component, MBP, has been injected into the experimental animals thereby
facilitating presentation to $CD4^+$ T cells. You would be quite right. The myelin basic
protein, as far as we know is not exposed to $CD4^+$ T cells and some event (possibly
infection) must occur that results in damage to the myelin sheath which results in it
being taken up by APCs for presentation to the $CD4^+$ T cells.

You will also realise, of course, that the actual damage caused to the nervous tissue is probably the result of a DTH reaction involving macrophage mediated damage although one cannot rule out other effector cells.

SAQ 8.12	We think it would be beneficial for you to make the conclusions from the data we have given you on the possible immune mechanisms involved in the protective effects of TCR peptides in EAE. Which of the following conclusions are supported by the data?

1) The protective effects are mediated solely by T cells.

2) The protective mechanism involves the elimination of encephalitogenic T cells.

3) The data suggests the involvement of idiotype-specific suppressor T cells.

4) The mechanism involves MHC I presentation of CD4$^+$ TCR peptides.

5) The data suggests that CD8$^+$ T cells are protective.

Immunotherapeutic approaches

The results obtained in investigations of EAE are creating intense interest because the lesions in EAE are very similar to those seen in patients with multiple sclerosis (MS). A recent report has shown that brain lesions from diseased patients with MS have lymphocytic infiltration and the T cells exhibit extremely limited Vβ gene usage in a similar fashion to that described for rodents. A similar homogeneity of the T cells causing the disease has been observed in human rheumatoid arthritis and experimental autoimmune diabetes in animals. This suggests that it should be possible in the not too distant future to at least reverse the deleterious effects of some autoimmune T cells in Man if suitable peptides based on TCR sequences of the offending T cells can be manufactured.

An alternative approach is to use the autoimmune T cells as a vaccine. Some workers have recovered T cells from joint fluid of arthritis patients or from spinal fluid of multiple sclerosis patients, grown them in culture (+ IL-2), inactivated them and then reinjected them back into the patients. Notice that this technique avoids determining TCR sequences or making peptides. We look forward to hearing of the results of these pioneering experiments.

Suppressor cell involvement

anti-idiotype suppressor cells may prevent autoimmunity

By now you will have examined the answer to SAQ 8.12 and found out that the autoimmune T cells may induce suppressor T cells with specificity for the peptide sequences found in the TCR chains associated with MHC Class I molecules. You may ask why exogenous peptides associate with MHC I molecules rather than MHC II. Well, there is some evidence that in activated T cells MHC I molecules actually recycle through the endosomes where they would pick up peptides derived from the proteolysis of endocytosed TCR and other surface molecules and present them on the cell surface. This combination would then be recognised by specific CD8$^+$ suppressor T cells which would then deactivate the cell.

8.10.4 Transplantation reactions

We have already looked into the mechanisms of allogeneic interactions and the basis of the CD4⁺ T and CD8⁺ T cell response. We shall now briefly examine the different mechanisms which may be involved in rejection of transplanted organs.

We have to say once again that we do not know a lot about the actual mechanisms involved in graft rejection. However, we do know that it is a mix of humoral and cell mediated immunity but with the tissue matching expertise available today, the major mechanisms appear to be mediated by cells.

Antibodies in transplantation reactions

antibodies can induce rejection, hyperacute rejection

Antibodies are known to mediate a very rapid type of rejection called hyperacute rejection. This type of rejection is initiated usually immediately after the host blood vessels are joined to the graft vessels and involves host antibodies binding to the endothelium initiating complement activation. This leads to platelet activation and aggregation leading to vascular occlusion (blocking of the vessel). This type of rejection is rare these days due to the careful matching of donor and recipient HLA and blood groups but used to be mainly due to antibodies directed against unmatched donor blood group antigens which are found on endothelial cells.

Antibodies have also been implicated in acute rejection which is complete by about 14 days after implantation and also involves complement activation by IgG antibodies specific for endothelial cell alloantigens. You will realise, of course, that NK cells could be involved using ADCC and macrophages and neutrophils using their Fc receptors.

∏   Before you read on, write down a list of the immune cells which you think are involved in transplant rejection and how they may act to destroy the transplanted organ.

Cells involved in transplantation reactions

T cells and rejection

There is good evidence to suggest that graft rejection is T cell mediated. In early experiments which lead to our knowledge of cell interactions, it was observed that thymectomised mice did not reject grafts and this also holds for the thymusless nu/nu mouse. Additionally, chickens, thymectomised at birth resulting in no antibody production, were found to reject skin grafts. More recent work has demonstrated that cloned alloreactive CD8⁺ T cell lines can adoptively transfer graft rejection.

allogeneic MHC I-specific CTLs

It is generally accepted that acute graft rejection is mediated mainly by T cells and transplanted kidneys undergoing rejection are found to have infiltrates of lymphocytes and macrophages. However, although donor-specific T cells are present in infiltrates of rejecting grafts some workers have found that they usually represent a minor portion of the total cells. The remainder of the cells include B cells, macrophages, NK cells and other cells. So we are still not sure what the mechanisms of tissue damage are. There is little doubt, however, as we have already indicated that allogeneic MHC I-specific CTLs are involved and lysis of renal parenchymal cells has now been demonstrated. The main targets in solid organs appear to be parenchymal cells and vascular endothelium in what we call acute rejection. Antibodies against alloantigens are also probably involved in this type of rejection which occurs in the first few week after surgery.

Events in T cell mediated rejection

passenger
leukocytes

The initiation of the transplant response is thought to be due to what are called donor derived passenger leukocytes or soluble donor alloantigens which find their way to the recipient's lymphoid tissue and activate alloantigen specific CD4$^+$ T cells. These can then activate CD8$^+$ T cells and these cells then find their way through the lymph and blood to the donor graft where they adhere to vascular endothelium using adhesion receptors and under the influence of chemotactic factors from some cells within the graft. They then migrate into the graft and initiate destruction by binding to donor alloantigen-bearing cells using TCR and cell adhesion molecules. Activation of CD4$^+$ T cells will result in cytokine production including chemotactic factors which attract monocytes from the blood. These will then become activated and secrete more inflammatory cytokines and mediate damage to the donor tissue along with CTL and possibly other cells such as NK cells.

chronic
rejection

There is also a very late rejection response called chronic rejection which can occur months after transplant surgery. The rejection organ is found to have a high degree of fibrosis and you could imagine that this response may parallel the sort of reactions involved in chronic DTH which lead to granuloma formation. Again we know little about the mechanism involved.

SAQ 8.13

Which one of the following statements is correct?

1) Chronic rejection involves granuloma formation.

2) The sensitisation step in graft rejection involves CD4$^+$ T cells being activated by alloantigens expressed on donor endothelium.

3) All transplantation rejection phenomena are controlled by CD4$^+$ T cells.

4) Hyperacute rejection is a result of antibody mediated damage by effector cells such as macrophages, neutrophils and NK cells.

5) Rejection processes are cytokine dependent.

8.11 Summary

You should now be able to appreciate the central role played by CD4$^+$ T cells and macrophages in cell mediated immune mechanisms and the evolving role of CD8$^+$ T cells as we learn that these cells may not be restricted to dealing just with virally infected cells and tumour cells. By now you should have also a good idea of the contribution made by cytokines and cell adhesion molecules to cell interactions in both CMI and humoral immunity. You will, we are sure, remember that most responses to antigen involves both humoral and cell mediated immunity.

We also dealt with tissue damage which is initiated by CD4$^+$ T cells and some interesting approaches to immunotherapy using experimental animal models. We are sure you notice the common mechanisms operative in CMI whether the CD4$^+$ T cells are being activated by cells infected with pathogens, allogeneic cells, tumour cells or cells presenting self peptides.

Objectives

Now that you have completed this chapter you should be able to:

- select a suitable protocol for separation of B cells, T cells, monocytes, polymorphs and red cells;

- describe methods for activation of T cells with mitogens;

- describe the experimental and theoretical basis of the mixed lymphocyte reaction and cell mediated lympholysis assay;

- predict the outcome of the MLR and CML based in the stimulating alloantigens;

- describe a cloning procedure for T cells;

- explain the involvement of T cells, macrophages, cytokines and cell adhesion molecules in the delayed hypersensitivity reaction;

- compare the mechanisms in innate immunity and DTH reactions;

- explain the possible involvement of $CD8^+$ T cells in antibacterial cell mediated immunity;

- list the conditions for differentiation of a $CD8^+$ T cell to a fully functional cytotoxic cell;

- describe the roles of $CD8^+$ and $CD4^+$ cytotoxic T cells in viral immunity;

- compare the killing mechanisms employed by natural killer cells, cytotoxic T cells and macrophages in tumour immunity;

- differentiate between Type I and Type IV hypersensitivity skin reactions;

- select procedures to identify intralesional T cells mediating tissue damage;

- formulate mechanisms based on given complex data in autoimmune reactions *in vivo*;

- identify the cells and mechanisms involved in transplantation reactions.

Responses to SAQs

Responses to Chapter 1 SAQs

1.1   The odd one out is 5), host lymphocytes, Items 1) to 4) are all antigens.

1) Donor cells are from another individual and bear a different set of self molecules which will be antigenic in the recipient.

2) Sperm carries the set of self molecules from the male which will be antigenic to the female.

3) Horse serum is antigenic to other species. There are species differences in many serum protein and these are recognised as nonself if injected into another species. You will notice the discussion on albumin later in the text.

4) Hen egg white lysozyme. Many animals produce lysozyme but again there are species differences and hen egg white lysozyme would be recognised as nonself in another animal.

5) Host lymphocytes in this context refer to an individual's own cells which are self and non-antigenic. This contrasts to the donor cells in 1).

1.2   This question was meant to further emphasise the ideas on self and nonself.

1) This involves the recognition of nonself. An early treatment for diphtheria was to give the patient serum from a horse immunised against *Corynebacterium diphtheriae* toxin. This serum contained antibodies against the toxin. The antibodies would block the activities of the toxin in the patient and also promote its destruction. However, as stated in SAQ 1.1, horse serum contains proteins including the anti-toxin antibodies which have somewhat different amino acid sequences to those in Man and these would be recognised as nonself and induce antibodies against the anti-toxin antibodies and other horse serum proteins.

2) This does not involve nonself recognition. Spermidine is part of the innate immune system and its production does not depend upon the recognition of the bacteria by the body.

3) C-reactive protein is part of the innate immune system and its production does not depend upon the body recognising the presence of a pathogen.

4) As with cowpox infections this involves recognition of nonself. It stimulates the production of a specific immune response.

5) This is quite a difficult question. If the plasma used in the transfusion contains different antigens from those in the plasma of the recipient, these will be recognised as nonself. The recipient's acquired immune system will be activated. If, however, the donor plasma contains identical antigens as those found in the recipient's plasma, then the recipient will accept the donor's plasma as though it was self.

1.3   1) Incorrect. They cannot discriminate between antigens and are not antigen specific.

2) Correct.

3) Correct. Proteins injected into another individual or another species may become nonself or antigenic.

4) Incorrect. Some serum proteins are part of the innate immune system eg acute phase proteins.

5) Incorrect. Some bacteria do not pass through the physical barriers ie enter the body, to be pathogenic. Examples include *Corynebacterium diphtheriae* and *Streptococcus spp* which develop colonies on the exterior of epithelial cells of the nasopharynx, pharynx and tonsils.

1.4   1) A conformational determinant (epitope) depends on the folding of protein and is lost on denaturation or proteolysis. Antibodies to the epitope can only bind the native antigen. So 1) is related to C).

2) Phagocytosis is promoted by complement activation particularly by the generation of C3b's which opsonise the pathogen thus promoting immune adherence to the phagocyte. 2) is related to D).

3) Epitopes are bound by antibodies. 3) is related to E).

4) A linear determinant (epitope) is a linear sequence of amino acids, so antibodies to these epitopes can bind to both native and denatured proteins. 4) is related to B).

5) Lysozyme attacks murein of bacterial cell walls. Thus 5) is related to A).

1.5   Answer 1)-4) all describe properties of environmental antigens. Answer 5) is incorrect as you can derive antibodies of more than one specificity to a single epitope.

1.6

1) Incorrect. Lymphoblasts are not the cells which produce antibodies for a few days and then die. The lymphoblasts give rise to plasma cells. It is the plasma cells which produce antibodies for a few days and then die.

2) Incorrect. Antibodies are indeed bifunctional molecules possessing both antigen binding and biological activities. However, the binding function is at the amino terminal ends and the other biological functions such as activation of complement at the carboxyl ends.

3) Correct. An antibody binds to the epitope because of non-covalent interactions between amino acids within the binding sites of the antibody and complementary sites on the epitope. In a particular experiment, 17 amino acids in each antibody binding site were found to interact with 16 residues on an antigen. You can imagine that, if you changed one of the residues on the antigen, the antibody would not lose its specificity but would bind with a different affinity ie it is cross-reactive.

4) Incorrect. The statement is saying that all the antibodies produced to a single antigen are monoclonal ie of one specificity. Since the antigen may bear many epitopes this is not correct.

1.7

Feature	T_H	T_C
Express MHC Class I	+	+ (1)
Recognises virally infected cells	-	+ (2)
Produce antibodies	-	- (3)
Binds MHC II	+	- (4)
Kills viruses	-	- (5)
Binds MHC I on virally infected cells	-	+
Binds cell associated antigens	+	+

Comments: 1) Since both types of T cells are nucleated, they both express MHC Class I. 2) At this stage in your studies you would be correct in choosing only cytotoxic T cells as recognising virally infected cells. However, as you will see later in your studies, T helper cells also recognise virally infected cells! 3) Although T helper cells promote the production of antibodies by B cells they do not synthesise antibodies. 4) T_H cells recognise antigen bound to MHC Class II on the antigen specific B cells. 5) Although cytotoxic T cells kill virally infected cells they have no mechanisms for killing the viruses.

1.8

1) The cells of transplanted kidneys invariably have a mismatch ie express a slightly different set of MHC I and II molecules to those of the recipient. These would be recognised as foreign and cells of the recipient would attack the donor cells and destroy the kidney. This does not occur due to the patient being given immunosuppressive drugs which prevent this response and the kidney continues to function in a 'slightly' allogeneic environment.

2) Based on the experiment described in 1.7.2 we could conclude that if a newborn was exposed to all the known MHC I and II molecules on mixtures of living donor cells, the newborn would become tolerant to them and would be able to accept transplants from any individual at some future time. Notice, we said

on a purely theoretical basis - it would not be an ethical procedure and there may be practical reasons why it would not work.

1.9

1) Incorrect. Although the alternative name for NK cells is large granular lymphocytes there is no convincing evidence to date that they develop from stem cells by this pathway.

2) Incorrect. These cells are first seen in the foetal yolk sac.

3) Incorrect. Although it is correct that T cells are the major cells in the recirculatory pool there are also memory B cells which form about 20% of the total cells.

4) Correct. Once B cells from the bone marrow have been activated by antigen in the peripheral lymphoid tissues some of the progeny become memory cells and enter the recirculatory pool.

5) Incorrect. The function of the bone marrow is the manufacture of blood cells ie the function of primary lymphoid organs and not the activation of T and B cells, a principal function of secondary lymphoid organs.

1.10

1) LGL. This matches with ADCC B) since this cell is the principal that performs antibody dependent cellular cytotoxity. It is not the only cell type that can do this since monocytes and neutrophils can also kill cells by this mechanism.

2) Neutrophils. This matches with phagocytic A) since phagocytosis is the principal activity of these cells. You will remember that neutrophils, like monocytes, cannot recognise antigen like T and B cells but they possess receptors for antigen coated in C3b's or antibodies. You may well argue that this answer could also be monocytes and you are quite correct. However, since each item can only be used once, you would have to use monocytes to match with Kupffer cells rather than phagocytic.

3) Monocytes. Blood monocytes are though to be the precursors for all tissue macrophages which include Kupffer cells E) found in the liver.

4) Memory cells. Since memory cells derive only from activated T and B cells they should be matched with antigen specific D).

5) Stem cells. These are only found in the bone marrow C).

1.11

1) This does not apply to lymph nodes since the statement is incorrect. The lymph is delivered to the node by afferent not efferent lymphatics.

2) This statement suitably describes a function of lymph nodes.

3) This does not apply to lymph nodes as the statement is incorrect.

   Although they do provide the environment for activating T cells, the majority of T cells enter the node direct from the blood and not by afferent lymphatics.

4) This does not apply to lymph nodes since the antigens are derived from the peripheral tissues and are carried by lymph to the node. This statement therefore should say that they trap lymph borne antigens.

Responses to Chapter 2 SAQs

2.1

This question was intended partly to help you assess your current knowledge of basic immunology.

1) True since they all move towards the anode. But, if you examine Figure 2.2 you will notice that antibody containing fractions, the serum gamma fraction and to a lesser extent, the β-globulin fraction move very little when subjected to electrophoresis. In contrast, albumin moves rapidly to the anode. We can conclude therefore that although they are all negatively charged, they carry different amounts of charge. Remember, however, that the charge on a protein is pH dependent. At very acidic pH's, all these molecules would carry a net positive charge.

2) False. Any antigen activates many different B cells to produce a wide spectrum of antibody specificities. We say the response is polyclonal, that is, a collection of many monoclonal antibodies are produced. This response is for two main reasons. Firstly, most antigens possess many different epitopes each capable of

eliciting an antibody of unique specificity. Secondly, even if there was only one epitope you would probably generate quite a few different antibodies as the B cells would 'see' the epitope from different angles.

3) False. Bence Jones proteins are not whole antibodies but light chains.

4) True. You can deduce this from Figure 2.2 where it shows that IgG comprises the major part of the antibody pool.

5) False. Even though such a blood transfusion would not elicit antibodies to the red blood cells, the blood contains other cells such as lymphocytes, platelets, monocytes and polymorphs all of which carry on their surfaces a unique combination of 'self' molecules of the Major Histocompatibility Complex. The recipient would have responded to these.

2.2   What you are looking for here are fragments that consist of essentially the F(ab')₂ components of antibodies. Since these fragments can bind 2 epitopes, they can link whole antigen molecules together to form a complex that comes out of solution. An examination of Figure 2.3 provides the answers to these questions.

1) No. Fab is only a single binding site and therefore cannot precipitate antigen.

2) No. When you treat IgG with 2-mercaptoethanol it cleaves the H-H and H-L disulphide bridges producing just heavy and light chains and these cannot precipitate antigen.

3) No. Fc possesses no antigen binding properties.

4) Yes. Pepsin attacks the heavy chains on the carboxyl sides of the H-H S-S bridges producing an F(ab')₂ fragment. This fragment does precipitate antigen.

5) No. If you examine Figure 2.3 you will see that this combined treatment results in Fabc fragments that lack light chains.

2.3   1) True. The molecular weight of pentameric IgM is about 900 000 D.

2) False. IgG and IgD represent the 4 chain monomeric unit; IgD is not a dimer whereas IgA (secretory IgA) is. IgM is indeed a pentamer. It is doubtful whether you can call IgE a member of the 4 chain monomeric unit club as it is unusual in having an extra domain similar to IgM in its heavy chains.

3) False. Variable domains are indeed found on both heavy and light chains. However, the second part of the statement is false as they are situated next to the first constant domain, not next to the hinge region.

4) False. This perhaps would mislead you if you did not consider all four subclasses of IgG. Although the H-L S-S link in IgG2, IgG3 and IgG4 is attached at the stated point, IgG1 is unique in having this S-S link connected to the carboxyl side of $C_\gamma 2$. This makes the statement false. Note C_γ is our specific nomenclature for the constant domains in the γ chain.

2.4   You should have been able to recognise structural characteristics in the right hand column that were unique to each of the members of the left column.

1) Hinge region. Match with E). One characteristic of this region is the high proline content.

2) Antigen binding site. Match with D) since the binding site consists of ($V_H V_L$).

3) Domain. Match with B) as a domain consists of about 110 amino acids.

4) IgM. Match with F) as IgM is the only polymer in the left hand column.

5) F(ab')₂. Match with C). Because this fragment consists of two linked antigen binding sites it may precipitate antigen.

6) Heavy chain. Match with A) as this represents an isotype - the structural characteristics that distinguish one antibody class from another.

2.5   This question was meant to help you learn the distinguishing features of IgG and IgM.

	IgG	IgM
1) 4 domains/heavy chain	+	(IgM has 5 - see Figure 2.8; do not forget the V domains)
2) Intersubunit S-S		+ (IgM is the only polymer)
3) Tail piece		+ (see Figure 2.8)
4) Kappa chains	+	+ (all Abs can have these)
5) J chain		+ (only polymers have this)
6) Pentavalent		+ (IgG is divalent)
7) Interheavy chain S-S	+	+

2.6   If you know all these you are doing very well!

	IgG	IgM	sIgA	IgD	IgE
1) No of subclasses	4	1	2	1	1
2) No of domains/mol.	12	70	24	12	14
3) Presence of J chain (+/-)	-	+	+	-	-
4) Heavy chain (Greek letter)	γ	μ	α	δ	ε
5) Molecular weight (kD)	146	900	380	180	190
6) Extended COO- end (+/-)	-	+	+	+	-
7) Monomers/molecule	1	5	2	1	1
8) K or λ chains/molecule	2	10	4	2	2

Notes: 2) Each monomer H chain has 4 domains, each L 2 domains making a total of 12. IgM and IgE possess an extra domain/heavy chain. 5) We inserted the molecular weight for IgD (180 000 D) as it was not in the text. Although it is a monomer like IgG it does have an extended hinge region (second paragraph of 2.2.4) resulting in a higher molecular weight. 6) You could have interpreted the extended COO⁻ region as representing the tail piece found in IgM and IgA. In addition to these antibodies, however, IgD has an octapeptide extension at the C-terminal ends of the heavy chains (see Section 2.2.8).

2.7   This question was intended to reinforce your understanding of antibody heterogeneity in terms of allotypes, isotypes and idiotypes.

1) False. Allotypes are in the constant domains of IgG.

2) True. Current thinking suggests that during an immune response, an individual produced anti-idiotype antibodies to the antigen specific antibodies. Therefore we would anticipate that other individuals could also do so. Antigen specific antibodies from Man injected into a rabbit can also elicit anti-idiotype antibodies.

3) False. Heterogeneity indicates differences in antibody structure. Since there are only 5 heavy chain isotypes plus some subclasses and two isotypes of light chains this does not contribute greatly to heterogeneity. Allotypes contribute more to heterogeneity since some subclasses of IgG possess a number of allotypes. The greatest contributor to heterogeneity, however, are idiotypes since each specific antibody possesses at least one idiotype and the body produces thousands of antibodies of different specificities.

4) False. Antibodies cannot be elicited unless the immune system is exposed to the antigen.

5) False. Since all classes of antibodies possess antigen binding sites it follows that all have idiotypes.

2.8   1) The antigen combining site is composed of one V_H domain and one V_L domain. Each domain contributes 3 **complementarity determining regions** that interact with an **epitope** on the antigen. The 3-dimensional configuration of the site is maintained by four **framework regions** in each domain.

2) Avidity is a measure of the binding strength of the **whole** antibody molecule and is estimated to be much hgiher than the **affinity** of a single antigen binding site.

3) The **variable** domains of an antibody can be **idiotypic**, resulting in the production of an anti-**idiotype** antibody. Each **idiotype** may consist of several **idiotopes** and the site recognised by the second antibody is called a **paratope**.

2.9   This question is to impress on you the importance of IgA. The clue is in the word respiratory that immediately suggests mucosal immunity.

1) No. Since the serum has normal levels of IgG we have to assume that he is producing antitoxins.

2) No. Although such a deficiency would have some effects, the levels of IgG would compensate.

3) No. This is a normal condition. Normal individuals have low serum levels of IgE.

4) This best fits the symptoms. If no secretory piece is produced then IgA cannot be localised on the respiratory mucosal surface and there is no protection against infections.

5) No. Since he has normal levels of serum IgA this is not true.

2.10   1) False. Although C3a does activate mast cells, it is not a chemotactic factor; that is a property of C5a.

2) False. Opsonisation also includes coating of antigen by antibodies leading to the complex becoming bound to phagocytes through the Fc receptors.

3) False. A high concentration of IgG is required to fulfil the requirement for 2 monomeric antibodies to be close enough together on, say, a pathogen, to bind C1q for activation. In contrast, a single IgM, since it possesses the equivalent of 5 IgG's in the polymeric structure, can activate complement by itself. It follows that you need much higher levels of IgG to do the same task that one IgM molecule fulfills.

4) False. Although this may be the impression from many texts, other non-antibody reagents can activate the classical pathway (see Figure 2.13).

5) True. Both classical and alternative pathways will only result in lysis of bacteria or cells in the presence of C5b's.

2.11   This is a very difficult question to assess the correct levels and you should be pleased if you have achieved above 50% correct. The results should, however, emphasise how the whole complement system is interdependent!

	Fc mediated phagocytosis	Lysis of bacteria	Chemotactic activity	C3b mediated phagocytosis
1) IgA-antigen complexes	+	+	+	+
2) Properdin deficiency	+	+/-	+/-	+/-
3) IgG deficiency	-	+/-	+/-	+/-
4) Mast cell malfunction	+/-	+	+/-	+/-
5) C4 deficiency	+	+/-	+/-	+/-
6) C5 deficiency	+	-	+/-	+/-

Notes: 1) IgA is not considered an opsonising antibody and does not fix complement conventionally so none of these functions would be adversely affected and they are all +. 2) Properdin deficiency would block the alternate pathway and therefore reduce complement activity but would not affect Fc mediated phagocytosis. It would, of course, have little effect on bacterial lysis and chemotactic activity of C3b-mediated phagocytosis when the classical pathway is activated. It would, however, cause significant reductions in these activities in innate immunity ie before antibody is elicited. 3) Since IgG is the antibody involved in Fc mediated phagocytosis, this would be absent. Additionally, since IgG is a major complement fixing antibody, the other activities would be significantly reduced. 4) Mast cells are the major source of chemotactic factors and malfunction of these cells would result in failure to attract phagocytes to the inflammatory site thus reducing phagocyte activity. However, the membrane attack complex should not be affected and lytic activity on bacteria would proceed normally. 5) C4 deficiency would block the classical pathway unless this was compensated by the alternate pathway. All activities would be effected except Fc mediated phagocytosis. However, as indicated previously this could be adversely affected due to lower numbers of phagocytes being attracted to the inflammatory site due to lack of C5a chemotactic factor and lower stimulation of mast cells. 6) C5 deficiency would have similar effects but would also block the alternate pathway resulting in total inhibition of the membrane attack complex and lysis of bacteria.

It would be a reasonable conclusion that C3b mediated phagocytosis would not be affected by either C4 or C5 deficiency. As stated above, however, there would be some reduction due to a lack of phagocytes being attracted to the site.

2.12   If you remember that the chemotherapeutic agent you are required to produce has to both block viral interaction with the cells and kill the virus then most of the alternatives presented to you are not valid.

1) If the treatment is longterm ie the patient needs several injections over a few weeks then a mouse monoclonal antibody will not be effective as the patient will probably produce antibodies against the mouse antibodies and they will be destroyed (see Section 2.2.4 on isotypes and 2.4.3 on complement).

   Additionally, this antibody may prevent virus infecting the cells but may result in destruction of the T cells. This would make the patient helpless to fight any infection and would not result in the destruction of the virus.

2) The only advantage of the human antibody over 1) is that it would not elicit anti-antibodies. Apart from that this treatment suffers from the same deficiencies.

3) The approach using the receptor is a positive one. The idea here is for the virus to bind the soluble receptor rather than the receptors on the cells so preventing infection. The attachment of F(ab')$_2$ to the receptor serves no useful purpose so we would expect this receptor to have a short half life giving only temporary protection or requiring lots of repeat injections. Additionally, this construct would not kill the virus.

4) A soluble receptor with a longer half life would be effective in preventing infection of the cells but would not result in virus destruction.

5) This is the best approach and the correct answer. By attaching IgG Fc to the soluble receptor you would probably endow the receptor with the long half life of IgG so it would give long term protection. In addition, IgG Fc would promote phagocytosis of the virus-receptor-Fc complex resulting in destruction of the virus. Approaches identical to this are presently being tried to eliminate the AIDS virus in infected patients.

6) Judging by the information given, a human antibody to the virus is not protective as all patients are seropositive ie they already possess antibodies and yet are not protected.

Responses to Chapter 3 SAQs

3.1   The best approach to this question is to establish the position of the band representing the *Eco*RI fragment of non-rearranged DNA ie that found in the sperm cell that hybridises with C$_\kappa$ mRNA. This is the 6kb band (sperm) and represents the fragments derived from both chromosomes (alleles). Any rearrangement of DNA would result in a change in the size of this fragment due to loss of *Eco*RI restriction sites arising as a result of the translocation of the constant gene closer to the variable genes. This could result in fragments that are larger or smaller than the 6kb band. A rearrangement in one chromosome would result in 2 bands, only one of which would differ in molecular weight from 6kb since only one chromosome has rearranged. Loss of the 6kb band and the appearance of 2 others would suggest that both chromosomes have rearranged. This then makes the answers quite straight forward.

1) This is obviously not true as any mature B cell has undergone rearrangement of DNA and is likely to have lost some of the germline DNA present in sperm cells. This, therefore cannot represent an explanation for any of the bands in the figure.

2) myeloma 1 pattern represents rearrangement on both chromosomes. This answer fits myeloma 1.

3) Rearrangement on one chromosome is represented by the myeloma 2 pattern since one of the bands is 6kb representing the unrearranged DNA on one of the chromosomes. This fits myeloma 2.

4) It is not possible to draw any such conclusion from the experiment. This response cannot be an explanation for any of the band positions.

5) This explains the appearance of one band with sperm.

3.2

1) Incorrect. Although the introns do comprise a major portion of the genome, it is the exons that encode protein products not the introns.

2) Incorrect. Variable and constant genes of heavy, kappa or lambda chains are always found on the same chromosome. For instance, the variable genes are found upstream of the constant genes for heavy chain on chromosome 14 (Man) and 12 (mouse) and so on.

3) Incorrect. VDJ joining only occurs in heavy chain genes as light chain genes do not have D segments.

4) Correct. During rearrangement, D-J joining results in deletion of any D segments downstream from the D segments involved in the D-J join. V to DJ joining result in the deletion of all remaining D segments and V segments downstream from the selected V segment.

5) Incorrect. There are only 2 V segments in mouse lambda chain genes. In Man there are more than 5 V segments (see Section 3.3.2 and Figure 3.5).

3.3

The answer is 4) = 0.008%. The extraordinary diversity of antibodies means that there must be more DNA committed to the expression of antibody genes than any other class of proteins. Some workers have estimated that we could probably get along quite happily with just 100 antibodies of different specificities. We thought it would be interesting for you to work out how much DNA it would need to encode the variable domains for 1000 antibodies.

Each V, D and J segment uses a triplet ie 3 nucleotides to encode each amino acid residue. In DNA this will be represented by 3bp. Thus using the information in the question we can calculate the number of base pairs for each gene segment. Thus for the V_H segments.

The 1000 x 98 x 3 = 294000bp

For the D regions 12 x 5 x 3 = 180bp

For the J regions 4 x 18 x 3 = 216bp

For the purposes of the calculation we can ignore the contribution by the D and J segments.

Then % of genome $= \dfrac{294\,000 \times 100}{3.5 \times 10^9} = 0.008\%$

3.4

This problem was to help you to sort out your V's, D's and J's in light and heavy chain genes, particularly the arrangements we find in lambda chain genes.

1) Not possible as the D-J joining has to precede the V-DJ joining . V-D arrangements do not ordinarily occur although recent evidence suggests it might (see Section 3.6.4).

2) Not possible as there are no D segments in lambda chain genes.

3) Not possible as $V_\lambda 2 J_\lambda 2$ would select $C_\lambda 2$ in the rearrangement. Additionally, $C_\lambda 4$ is a pseudogene (see Figure 3.4).

4) This is a possible murine rearrangement where D2 joins with $J_H 3$ and then pairs with $V_H 4$.

5) Not possible as there is only one constant gene for kappa chains.

3.5

This question should have helped you sort out the 12/23bp rule.

1) Incorrect. Although heptamers and nonamers are conserved, spacers are not.

2) Incorrect. V_H gene formation requires first, a D-J rearrangement followed by a V to D-J rearrangement ie 2 DNA rearrangements.

3) Incorrect. Although the statement is true for heavy chain genes where both V_H and J_H segments have 23bp spacers, it is not correct for light chain genes. In lambda genes the V segments have 3' 23bp spacers and the J segments have 5' 12bp spacers; in kappa genes the converse holds. Since the 12/23bp rule is adhered to in light chain genes V segments can join with J segments.

4) Correct. V_H does encode framework regions 1-3 and CDR 1/2.

5) Incorrect. The heptameric sequences are deleted along with all other intervening DNA between the two joining segments.

3.6   You should not have had too much difficulty with this one!

1) Each V segment encodes the last 3 amino acids of the leader exon and the first 95 or so amino acids of the variable domain including 2 CDRs and 3 framework regions.

2) Joining mechanisms involve the complementary base pairing of **heptameric** and nonameric sequences which are recognised by a **recombinase** enzyme.

3) V segments in lambda chain genes have a **23** nucleotide spacer and the J segment a **12** nucleotide spacer. The arrangement on the chromosome downstream from each V segment is a **heptamer** followed by a spacer then a **nonamer**.

3.7   1) Incorrect. Junctional diversity can occur at any joints within CDR3 ie D-J, V to DJ in heavy chain genes and V-J in light chain genes.

2) Correct. Any joining resulting in frame triplets TAA, TGA or TAG will encode stop or termination codons.

3) Incorrect. Not all joining results in junctional diversity.

4) Incorrect. Junctional diversity does not always involve V joining but may also occur during D-J joining.

5) Incorrect. N segments are only inserted in heavy chain rearrangements.

3.8   The correct answer is combination e). This was probably the most difficult problem in this chapter so do not be too concerned if you did not get the right combination. However, it was a useful exercise because we have been discussing detection of mutations and you may have been asking how they could be distinguished from junctional diversity and the like.

From Figure 3.9 the V3' flanking heptamer is CACAGTG and the J5' flanking heptamer is CACTGTG. Having identified these, you could organise yourself by working out the amino acid triplet codons;

V3' -	TTG	TAC	CCT	GTC	AGA	ACC	**ACA**	**GTG**	TTT	ACT	
	V1	V2	V3	V4	V5						
	leu	tyr	pro	val	arg						
	ACT	**CAC**	**TGT**	GCA	TAC	TGC	TAC	TTC	GAT	- J5'	
					J5	J4	J3	J2	J1		
					tyr	cys	tyr	phe	asp		

You can see that the joining in the absence of junctional diversity occurs when a V3' codon joins directly to a J5' codon.

Now examine each sequence:

1. You will notice that ser has replaced the arg encoded by V5 (AGA). From the Table ser is encoded by AGT or AGC ie a single base change. Since there has been no other amino acid changes in the sequence this is due to mutation and not to junctional diversity.

2. Again, the arg has been replaced by ser but there has also been a reduction of amino acids to 9 in the sequence. The actual sequence change (emboldened) is leu-tyr-pro-val- **ser-cys** -tyr-phe-asp. This means that tyr (J5) has been deleted in the join. Based on this knowledge we can see where the join was made. Examine the proximal V and J triplets, You will notice that the join was made between the GA of AGA and AC of TAC

resulting in a new codon of AGC and a reduction of 3 nucleotides. The sequence would therefore still be in frame. Sequence 2 therefore is due to junctional diversity.

V5	J5	J4	J3	J2	J1
arg	tyr	cys	tyr	phe	asp
AGA	TAC	TGC	TAC	TTC	GAT
AGC		TGC	TAC	TTC	GAT
ser		cys	tyr	phe	asp

3. This sequence results from an insertion of an extra amino acid between arg (V5) and tyr (J5). The extra triplet must derive from the sequence 3′ of V5 and 5′ of J5. The inserted amino acid thr is encoded by AC(X) where X is any of the 4 bases. Examine the sequences:

AGA ACC ACT TGT GCA TAC

V5   J5

There are two possibilities here:

• Since ACC encodes thr the DNA was cut after ACC and 5′ of J5 to give AGA ACC TAC ie arg-thr-tyr.

• The cuts were after the middle C of ACC and after C of GCA to give the sequence AGA ACA TAC ie -arg-thr-tyr- the same result. Both of these are due to junctional diversity.

4. In this sequence the change occurs in V2 from tyr to phe and therefore not within the V-J join. Notice the sequence apart from this amino acid change is identical to that seen with no junctional diversity. A single base change from TAC to TTC would account for the switch to phe. This is a mutational change.

We have already shown 2 mutations in sequences 1 and 4. The answer to this SAQ is therefore e).

5. We might as well do the final sequence. Notice that there is an ala inserted between V5 (arg) and J5 (tyr). As in 3 we have to examine the sequences between the triplet codons for these amino acids.

AGA ACC ACG TGT GCA TAC

V5   J5

Since ala is encoded by GC(X) the DNA was cut immediately 3′ of V5 and 5′ of GCA leading to the sequence AGA GCA TAC ie -arg-ala-tyr-. We have another case of junctional diversity.

This exercise should impress you with the potential of junctional diversity to produce major changes within the CDR3.

3.9

1) Somatic mutation is matched with e) (Immune response) since this occurs during responses to antigenic challenge.

2) CACTGTG matches with a) (V_H exchange) as it appears to be this embedded heptamer that promotes exchange of a 5′ V_H segment for the V segment in a previously rearranged VDJ.

3) $CD5^+$ B cells match with d) (germline genes) as these B cells appear to use V_H segments in unmutated form.

4) Junctional diversity matches with c) (CDR3) as it is imprecise joining within CDR3.

5) Rheumatoid factor matches with b) (Fc region) as this polyreactive antibody possesses specificity for the Fc region of IgG.

3.10

1) Incorrect. Class switching does result in the same idiotype, ie the same antibody specificity and a different isotype. But, since inherited allotypes are carried on heavy chain constant domains and are unique to Ig classes or subclasses, these would be different.

2) Incorrect. Alternative RNA splicing involves the production of long mRNA transcripts encompassing both constant clusters involved in the switch and does not involve deletion of DNA. It therefore does not involve a loss of constant genes.

3) Correct. Since the IgG1 constant gene is downstream from IgM and IgG4 gene is downstream from that of IgG1, these isotypes could be produced in this sequence.

4) Correct. Whereas secreted IgM possesses an additional tailpiece of 20 amino acids on the carboxyl end of $C_\mu 4$, membrane IgM possesses a transmembrane portion of 26 amino acids and a cytoplasmic tail of 3 amino acids.

5) Incorrect. All except IgD possess switch regions (look back at Figure 3.12 and the associated text).

3.11   This was intended to further demonstrate the manipulation of restriction sites in vectors.

1) Incorrect. This was already demonstrated in Figure 3.15. The EcoRI-HindIII fragment containing the murine heavy chain variable region together with the HindIII-BamHI fragment containing the human IgM constant region can be introduced into pSV2neo by ligation into the EcoRI-BamHI site.

2) Incorrect. The HindIII-HindIII fragment containing the murine light chain variable region together with the HindIII-BamHI fragment containing the human kappa constant region can indeed be ligated into the HindIII-BamHI site in pBr322. However, this cannot be used to infect mammalian cells since the vector does not contain the elements to promote transcription or polyadenylation sequences for expression in mammalian cells.

3) Incorrect. You must generate complementary sticky ends for successful ligation. Although some restriction endonucleases do indeed produce the identical sticky ends (these are called isochizomers) that is not the case in the range of enzymes given. Since the variable region of the murine light chain is contained within a HindIII-HindIII restriction fragment it cannot be inserted into pSV2neo which contain EcoRI, ClaI and BamHI restriction sites only.

4) Incorrect. Although the first part of the statement is true, it is possible to ligate the light chain restriction fragments into pSV2neo in a round about way and subsequently to transfect SP2/0. See answer 5).

5) Correct. We are agreed that the chimaeric heavy chain segments can be successfully ligated into pSV2neo. We have also concluded in 3) that we cannot directly ligate the light chain segments into pSV2neo. But you could do it in two steps. First ligate the chimaeric light chain segments into pBr322 as we concluded in 2). Notice that there is a ClaI restriction site next to the HindIII site. You can now remove the slightly longer ClaI-BamHI fragment from pBr322 and ligate it into the ClaI-BamHI restriction site in pSV2neo. This can be used to transfect SP2/0 together with pSV2neo carrying the chimaeric heavy chain genes and both can be expressed resulting in IgM antibodies.

Responses to Chapter 4 SAQs

4.1   This is a difficult question but the answers provide a lot of basic information about the immune response.

1) Incorrect. Since the carrier is immunogenic on its own we would expect B cells to respond to carrier determinants as well. We did not detect these antibodies in the experiment reported in Figure 4.1 because we are only examining an anti-hapten response.

2) Incorrect. Injection of hapten carrier 3 into the recipient mice would result in priming of T cells and B cells which recognise C3 resulting in a primary response. We can call such cells C3 T cells and C3 B cells.

3) Correct. Recipient mice would not give a secondary response to H-C2 since C2 is not covalently attached to the hapten. However, since the mouse receives H-C3, T cells in the recipient mouse would become primed to C3. If these cells are transferred to another mouse and this mouse is given H-C3 a secondary response would result because C3-primed T cells and hapten-primed B cells (from Donor 1) are present.

4) Incorrect. Since the carrier C2 primed T cells are destroyed using the anti-T antiserum, there would be no secondary anti-hapten response.

5) Incorrect. Destroying C1-primed T cells has no effect on the response to H-C2.

4.2   1) Antigen presenting cells prime T cells c).

2) Polymorphism is the possession of many alleles d) at a single locus.

3) Homozygous should be matched with same alleles a) as the alleles at any one genetic locus are the same in a homozygous individual.

4) Allogeneic describes non-identical members of the same species e).

5) MHC should be matched with b) as it is central to recognition of self.

4.3

1) At the **centromeric** end of the chromosome in Man we find a cluster of genes called **DP, DQ** and **DR**. These are expressed on **antigen presenting** cells and activate **helper** T cells.

2) In mice, the K and D loci regulate **cytotoxic** T cells. They also induce **graft rejection** when tissues are transplanted into **allogeneic** mice.

3) **Class** I genes in Man are found at the **telomeric** end of the MHC complex. Because of the many **alleles** at single **loci**, we say they are very **polymorphic**.

4.4

This question should have helped you understand the relationship of haplotypes with recognition of self.

1) Incorrect. Since the child's cells would express Class I molecules B12, B27, A2 and A29, these include the Class I molecules expressed on the father's cells ie B12 and A29 and recognised as self by the father's cytotoxic T cells. Therefore these cytotoxic T cells would also recognise the child's virally infected cells as self if exposed to them (say in a culture dish) irrespective of the presence of the allogeneic B27 and A2 molecules. This will become clearer later in the text.

2) Correct. The cells of the father would recognise maternal alloantigens on the cells of the child as allogeneic and would attack the cells.

3) Correct. In this case, it would be a host versus graft situation. The cells of the child would recognise the maternal alloantigens expressed by the haplotype the child did not inherit ie B7 and A28.

4) Incorrect. Since the cells of child and father share Class II DR4, the T helper cells of the former would provide help if the cells were mixed in cell culture.

5) Incorrect. We have already said that the maternal cells would be destroyed in 3).

4.5

We think you would have found the majority of these quite straightforward. Class III code for complement proteins which are part of innate immunity (A); all 3 classes of genes exhibit polymorphism (B) and Class III products are soluble products secreted from the cell whereas both Class I and II gene products are expressed on the cell surface (C).

Characteristic	Class I	Class II	Class III
Present on all nucleated cells	+	-	-
3 extracellular domains	+	-	-
Chains of 44kD	+	-	-
β_2-microglobulin	+	-	-
Innate immunity connection	-	-	+ (A)
Stimulate helper T cells	-	+	-
Bind peptides	+	+	-
Target for cytotoxic T cells	+	-	-
Polymorphism	+	+	+ (B)
$\alpha\beta$ gene product	-	+	-
Soluble products	-	-	+ (C)

4.6

1) Incorrect. This congenic strain is indeed the correct haplotype since it is the convention to write the mouse strain providing the background followed by that providing the MHC. However, the C57BR was providing the MHC and the B10 the background.

2) Incorrect. The experiment described in Figure 4.8 informs us that there are two distinct populations of T cells for each parental haplotype in F1 mice. In this statement, therefore, there would be T cells which provide help to B cells expressing CBA (H-2^k) molecules and different T cells for B cells expressing C57BL/10 (H-2^b) molecules. Since these T cells were transferred into a congenic mouse strain in which all B cells only express B10 (H-2^b) molecules, the T cells recognising CBA (H-2^k) molecules + SRC would not be activated and they would reappear in the TDL.

4.7   It is advisable to draw yourself a diagram so you know which haplotypes are where or you will get confused. This question is difficult. If you got it right, you have learned a lot about thymic education and you can congratulate yourself.

1) Incorrect. It is true that the recipient and donor cells are allogeneic for MHC Class II. However, we have emphasised that the phenotype of the T cell has nothing to do with the ability to collaborate in a chimaera. The important point is what the T cells now recognise as self; this is obviously the k haplotype of the recipient mice. Now what B cells are present in these mice? Only those from the donor and these express MHC II of the d haplotype of the donor. So the T cells which are syngeneic to the B cells cannot collaborate because they no longer see the d haplotype as self. Although 1) states that there would be no cell co-operation, it is not because the recipient cells and donor cells are allogeneic for MHC Class II but there are no recipient B cells - that would be recognised as self after T cell maturation in the recipient and thus donor B and T cells can no longer co-operate.

2) Incorrect. Although there would be no collaboration, it is possible that B cells could respond to T-independent antigens. We did mention much earlier that some antigens were independent of T cells. These T-independent antigens can activate B cells directly and do not require cell interaction between T helper cells and B cells.

3) Incorrect. The basis for this is the same as 1). Donor B and T cells cannot collaborate!

4) Incorrect. This is simply writing 3) in another way since donor B cells are d haplotype and T cells recognise the k haplotype.

5) Correct. For the reasons stated in 1) there would be no collaboration.

4.8   Did you manage to think this one out? The answer is really quite straightforward if you establish a couple of observations before you start. Firstly, irradiated recipients do not have any lymphoid cells, neither T nor B cells. Secondly, any response is due to donor T and B cells. Since the donor T cells of strain A have adaptively differentiated in the B strain mouse to recognise strain B MHC II these cells will only provide help to cells expressing these molecules. However, the donor B cells still express strain A MHC II, therefore the T cells cannot provide help and no antibody results.

4.9   1) Correct. Exogenous antigens pass through the endosomal pathway.

2) Correct. Endogenous antigens are converted into peptides in the cytoplasm eg the endoplasmic reticulum, become bound to MHC I molecules and are presented to CD8 cells.

3) Correct. Since an F1 mouse possesses two haplotypes there must be two sets of T helper cells since a single T cell learns to recognise a single haplotype as self.

4) Incorrect. Although the statement correctly describes MHC restriction, there are no B cells expressing the B haplotype as they were destroyed by irradiation (see also SAQ 4.7). Remember that the A strain T cells matured in a B strain mouse would have 'learned' to recognise the B haplotype.

5) Correct. Antigen must be subjected to proteolysis before being presented.

4.10   This question simply emphasises the existence of amino acid residues which interact with either the TCR or MHC. Statement 1) tells us that amino acids 55 and 56 are involved in either interaction with the TCR or MHC; no other conclusions can be drawn. However, since alanine (56) mutant binds to the MHC (statement 2)) and must be presented, the failure must be in T recognition and residue 56 is a TCR determinant. Statement 2) also tells us that alanine (53) mutant binds to the MHC but statement 4) indicates a failure to stimulate the T cells. Alanine 53 is therefore a TCR determinant. The fact that alanine (55) mutant fails to compete with native HEL peptide and fails to stimulate T cells (statement 1)) indicates that residue 55 is essential for MHC binding.

Only statement 3) is irrelevant to these conclusions so the answer is 1), 2), 4) and 5).

4.11   This should have helped you to learn some of the differences and similarities between exogenous and endogenous processing.

	MHC I	MHC II
Binds self peptides	+	+
Exogenous antigens	-	+
Involves invariant chain	-	+
Proteases	+	+ 1)
ER to Golgi pathway	+	+ 2)
Late endosome	-	+
Peptide binding cleft	+	+

Comments: 1) Obviously both pathways involve peptides since this is the form that binds MHC and proteases degrade both types of antigens. 2) Since both MHC I and II molecules are synthesised in the ER and then pass to the Golgi this is a common pathway before they diverge, the MHC II molecules being transported to early or late endosomes.

4.12   2) is correct. This was difficult. If you could not do it, do not be too concerned. If you gave the correct answer for the right reasons, you did very well!

In this experiment we are examining the education of cytotoxic T cells and T helper cells in the thymus and you have to find a reason for the apparent failure of the thymus to educate the incoming T cells to recognise self. For instance, A strain cells matured in an F1 mouse appear to have failed to learn that the B haplotype is self. We would have anticipated that target cells of both A and B haplotype would have lysed; do you agree? A similar argument holds for the other results. What we see however is the lysis of cells that bear the haplotype of the parent cells (BM cells) only.

Firstly, let us say that the pre-T cells did learn to recognise the MHC I and II haplotype in the recipient thymus as self. For instance, pre-T cells from A strain mice learned to recognise haplotype A or B in the F1 recipient. These cells, however, only lysed A targets ie cells of the parental haplotype.

The reason for this is a total loss of antigen presenting cells of the host haplotype due to irradiation. The only APCs available then would be those which have developed from the bone marrow cells of parental haplotype (A). As we said earlier, development of virus specific cytotoxic T cells depends on IL-2 provided by T helper cells. The latter need to be activated by MHC II + peptide on APCs in order to release IL-2. As we have just said, the only APCs available only express parental haplotype (A) thus activating helper T cells which recognise MHC II (A) + viral peptide. Since the IL-2 acts at short distances we would expect that cytotoxic T cells that interact with MHC I + peptide on the same APCs will be stimulated. This would result in the activation of cytotoxic T cells with specificity for A + peptide only, hence the result.

You should be able to argue similarly for the other 3 experimental observations in the table.

Responses to Chapter 5 SAQs

5.1   The most likely reason for the failure to detect any bands on Western blotting of the TCR extract is that the monoclonal antibodies, which successfully precipitated the TCR from cell lysates, did not bind to the TCR on the nitrocellulose membranes (answer 2)). Since monoclonal antibodies are directed against a single epitope, it is often the case that they are not useful in Western blotting. The reason for this is that during the processing in the SDS buffer, the proteins are denatured and the epitope is no longer in its native form. The monoclonal antibody can no longer bind to the epitope.

5.2   We expect this did not take you too long if you have learned the basic features of the TCR.

Structural features	α-chain	β-chain
Binds peptide	+	+
Acidic glycoprotein	+	-
Homology to Ig	+	+
CD8$^+$ cells	+	+
Transmembrane lysine + arginine	+	-
CD4$^+$ cells	+	+
V, D and J exons	-	+
Binds MHC	+	+

5.3 This question ensured that you knew the major differences between the two types of receptors although it has to be admitted we are still ignorant of some aspects of the TCR-1.

Feature	TCR-1	TCR-2
Antigen specific	+	+
CD8$^+$	-	+
Heterodimer	+	+
MHC restricted	+/-	+
CD4$^+$	-	+
CD3$^+$	+	+
Restricted specificity	+	-
Most common receptor	-	+

Whichever answer you gave to MHC restricted (TCR-1) could be right. As we said in the text, some workers have shown restriction, others have shown non-restriction! The restricted specificity obviously refers to the narrow range of antigens that TCR-1 bearing cells are active against.

5.4
1) Incorrect. Although it is true that the β chain locus is the one to have been completely sequenced, it is the α chain locus which has more than 50 J segments (see Figure 5.5).

2) Incorrect. N region additions do not occur in light Ig (L) chain gene rearrangements.

3) Correct. Because the δ chain locus is situated between the variable and joining segments of the α chain locus, rearrangement of the α chain genes would eliminate the δ locus.

4) Correct.

5.5 A relatively simple question but you do need to know these phenotypic differences. These demonstrate some of the markers on all T cells and those on either of the two populations.

Surface molecule	T helper cell	T cytotoxic cell
TCR	+	+
CD3	+	+
CD4	+	-
CD8	-	+
CD2	+	+

5.6 Again, it is essential that you know these interactions that promote adhesion and activation.

1) CD11aCD18 (LFA-1) matches f) ICAM-1.

2) LFA-3 matches a) sheep red cell receptor (CD2).

3) Collagen matches g) CD26.

4) CD28 matches b) B7 (BB1).

5) VLA-4 matches d) Fibronectin and VCAM-1.

6) CD4 matches c) MHC II.

7) MHC I matches e) CD8.

5.7

1) False. Positive selection is the process whereby TCR-1 and TCR-2 expressing cells with any affinity for MHC are rescued from cell death. The MHC is presented on thymic epithelium not bone marrow derived cells.

2) False. Although true for TCR-2 expressing cells, TCR-1 cells generally do not express CD4 or CD8 and gene rearrangement therefore does not precede a non-event!

3) True. Foreign antigen is not involved in the selection. We think development is selected for by MHC + self peptides.

4) True. For instance, T cells using the V_β 17a chain recognise I-E products on B cells but not on macrophages. The conclusion was that the peptides expressed by the B cells associated with I-E were different to those on macrophage. In case you are confused, the V_β 17a simply represents a particular V exon the product of which was identified by antibodies.

5) False. Although these are differences, you also know that TCR-1 expressing cells do not normally express CD4 or CD8.

5.8

Having completed this question you will understand the principle behind the panning technique. The anti-Ig will remove B cells from a cell population which has not been treated previously with any antibodies. The anti-Ig binds the B cell antibody receptors. However, if cells have been previously incubated with an anti-marker antibody then the cells bearing the marker will also be retained on the anti-Ig plate. The best preparation of cytotoxic T cells would be derived from protocol 1).

1) The first stage causes adherence of B cells; the nonadherent cells (all the T cells) are then treated with anti-CD4 and placed on an anti-Ig plate. This would remove helper T cells. The nonadherent cells would be enriched cytotoxic/suppressor T cells.

2) The adherent cells resulting from the first stage would be helper T cells ($CD4^+$). Subsequent treatment of the non-adherent cells with anti-$CD8^+$, therefore, would result in both T cytotoxic cells and B cells binding to the second anti-Ig plate. Thus the adherent cells would be a mixture of B and T cells.

3) As for 1), the first stage would result in nonadherent T cells. Subsequent treatment with anti-$CD8^+$ would result in cytotoxic/suppressor T cells being coated. However, since the treated cells were placed on an anti-CD4 plate, the adherent cells would be helper T cells.

4) Anti-CD7 would bind all T cells as CD7 is a pan-T marker. Treating the nonadherent cells with anti-CD8 would have no effect as no T cells are present. Therefore, exposing the nonadherent cells (B cells) to an anti-Ig plate would result in adherence by B cells.

5) Treatment of cells with anti-CD2 followed by binding to an anti-Ig plate would remove all T cells as CD2 is on all T cells. However, the plate would also bind all B cells. The adherent cells, therefore, would include all the original cells. When exposed to an anti-CD4 plate, the nonadherent cells would include cytotoxic/suppressor T cells and B cells. Hence this protocol does not result in an as enriched cytotoxic T population as 1).

5.9

1) Incorrect. All except CD5 are present on all B cells. CD5 is present on the subset of B cells which produce polyreactive antibodies.

2) Incorrect. ICAM-1 is considered to simply promote adhesion and is not directly involved in activating T cells. The remainder do promote adhesion and activation.

3) Incorrect. The enzyme-linked immunosorbent assay measures the amount of antibodies present not B cell proliferation.

4) Incorrect. The hemolytic plaque assay can assess the number of antibody secreting B cells, not the amount of antibody.

5) Correct. Most B cells express CD21 which is the receptor for EBV.

5.10 If you got these right you have demonstrated that you know something about the characteristics of APCs.

Function/phenotype	B	T	La	IDC	FDC
Classical APC	+		+	+	
FcR⁺, CR⁺, MHC I/II⁺			+		
Antigen specific	+	+			
FcR⁺, CR⁺, MHC I⁺II⁻					+
Iccosome production					+
Activates unprimed T cells				+	
Activates primed T cells	+		+	+	
Activates primed B cells		+			+

Comments: Most of these are straight forward; we were obviously looking for cell types which all expressed the characteristic. A majority of B cells express FcR and some CR in contrast to all Langerhans cells. Similarly some T cells express FcR, CR and MHC I but all FDCs express this phenotype.

Only IDC activates unprimed T cells but classical APCs, (those expressing MHC Class II ie B cells, Langerhans cells and IDC), can activate primed T cells. B cells are activated by helper T cells but primed B cells are also activated by FDC, with cytokine help from T cells.

5.11

1) Incorrect. B cells do express idiotypes on their receptors but memory cells do not express IgM.

2) Incorrect. B memory cells are part of the recirculatory pool but express high levels of L-selectin which is probably necessary for them to exit from the blood stream.

3) Incorrect. B memory cells have undergone somatic mutation but do not express high affinity IgM or IgD, only the other isotypes.

4) Correct. It is thought that memory cells, after undergoing somatic mutation to a high affinity cell are selected for expansion by antigen on FDC. Cells which fail to do this, most likely the majority, are thought to die.

5) Incorrect. CD5 B cells are not thought to undergo somatic mutation but maintain expression of germline immunoglobulin genes - they presumably are not memory cells.

5.12

Property	Helper T cells	B cells
MHC II restricted	+	-
Interacts with FDC	-	+
Primed by all APCs	-	- (1)
Expresses MHC II	-	+
Zeta/eta chains	+	- (2)
Positive selection	+	- (3)
Expresses MHC I	+	+
Cytokine production	+	- (4)
Native antigen	-	+ (5)
Haemagglutination assay	-	+ (6)
Memory cells	+	+
CD45⁺ LFA1⁺	+	+
Homing receptors	+	+

Comments: (1) T cells are primed by IDC only. (2) zeta and eta chains are part of the CD3 complex. (3) Positive selection is the process by which MHC binding T cells are selected in the thymic cortex. (4) Cytokine production is one of the principle features of T cells. (5) B cells bind antigen in the native state, T cells see only linear peptides. (6) The haemagglutination assay can be used to detect the product antibodies by B cells.

5.13 This question gives the basis for comparison of T helper and T suppressor cell characteristics. 2-5 are common to both.

Function	T helper	T suppressor
1) They bind antigen alone	-	+
2) They may be MHC II restricted	+	+
3) They may be idiotype specific	+	+
4) They secrete regulatory molecules	+	+
5) They are derived from CD4$^+$CD8$^+$ cells	+	+

Comments: They are very similar, aren't they. There is some evidence that T suppressor cells bind antigen alone; whether they all do is not known. They both express idiotypes by virtue of the fact that they are both antigen specific. T suppressor cells secrete suppressor factor, T helper cells a wide range of lymphokines. All T cells early in development express both CD4 and CD8; later T helper cells lose CD8 and T suppressor cells CD4 as they mature in the thymus.

Responses to Chapter 6 SAQs

6.1 1) Correct. Some antibodies to CD3 activate all T cells in an antigen nonspecific manner and hence the T cells would produce cytokines.

2) Incorrect. Although these procedures involved a lot of work, they rarely resulted in the purification of a cytokine.

3) Correct. Mitogens activate a majority of T cells in an antigen nonspecific manner resulting in cytokine production.

4) Correct. In most instances, the mother will be semi-allogeneic to the child and the child will also express paternal MHC molecules. Hence there would be two way stimulation of T cells in what we call a mixed lymphocyte reaction resulting in production of IL-2.

5) Correct. Since the supernatant may contain a mixture of cytokines, some inhibitory to the action of others, it is possible not to be able to detect some cytokines in unfractionated supernatants.

6.2 1) Paracrine matches with neighbours c) since cytokines acting on neighbouring cells are acting in a paracrine fashion.

2) Pleiotropic matches with many cell targets e) since this term indicates a cytokine acting on many different cell types.

3) Interleukin 2 matches with receptor a) since it acts through a membrane receptor.

4) Autocrine matches with single cell b) since this term means a cytokine produced by and acting on the same cell.

5) Antibodies match with concentrate cytokine d) since they are used to purify and concentrate cytokines from supernatants.

6.3 1) Incorrect. IL-2 is 133 amino acids, the immature product has a 20 amino acid signal peptide. The remaining statement is correct.

2) Incorrect. NK cell activity is only augmented by high concentrations of IL-2. (Note that we will learn that this is presumably due to the fact that NK cells do not produce IL-2R$_\alpha$ chains).

3) Correct.

4) Incorrect. It is generally accepted that all $CD8^+$ cells do not produce IL-2. We will examine this again in Chapter 8.

5) Correct. If you remember the report on costimulator, the cells were only IL-2 dependent at low cell concentrations. It is thought that there are not enough IL-2 producing cells to support growth at these concentrations.

6.4 This was a useful exercise to enable you to see how to calculate the numbers of any type of receptors on cells. Did you succeed? Here is how you do the calculation.

First, you can find out the numbers of picomols of IL-2 attached to the cells. You simply divide the counts bound to the cells by the specific activity of the radiolabel.

No of picomols attached to cells $= \dfrac{4500}{1.7 \times 10^6}$ picomols and this is $\dfrac{4500}{1.7 \times 10^6} \times 10^{-12}$ mol.

Since the number of molecules of IL-2 per mol is Avagadros number

then the number of IL-2 molecules attached to the cells is:

$$\frac{4500}{1.7 \times 10^6} \times 10^{-12} \times 6 \times 10^{23}$$

If you divide this result by the number of cells you obtain the number of receptors per cell

$$= \frac{\dfrac{4500}{1.7 \times 10^6} \times 10^{-12} \times 6 \times 10^{23}}{4.5 \times 10^5}$$

$= 3529$ IL-2 receptors per cell

6.5 If you get provided all the right explanations for the observations you are on your way to becoming a good practical immunologist!

1) The anti-Tac would only precipitate the IL-2 cross linked to the Tac receptor (p55) resulting in the 68kD band. However, you will have noticed that some IL-2 bind receptors but are not affinity linked. The IL-2 would remain bound to the Tac during the precipitation with anti-Tac but the complex falls apart in the SDS buffer. You would therefore observe a 15kD band due to Tac-bound but not affinity linked IL-2.

2) The presence of anti-Tac during affinity labelling would prevent IL-2 binding to Tac and hence this radiolabelled band would be missing in this experiment.

3) Picomolar quantities of IL-2 mainly promotes intermediate affinity linking ie to the p75 resulting in the 90kD band. For the same reasons as 1) you would also see a 15kD band.

4) Nanomolar quantities of IL-2 promotes affinity linking of both intermediate and low affinity receptors, thus resulting in binding of IL-2 to both p75 and Tac.

5) Excess non-radiolabelled IL-2 would dilute out the radiolabelled IL-2 so that very few receptors would bind the radiolabelled IL-2. Most would be binding the unlabelled IL-2. An autoradiogram would therefore show no radioactive bands. However, if the protein bands are transferred to a nitrocellulose membrane and a Western blot performed using a polyclonal anti-IL-2 you would detect all three bands since this would detect the IL-2R chains affinity labelled with the nonradiolabelled IL-2. Get the idea?

6.6 This was quite straightforward and simply helps you to distinguish between the structural and biological features of the two IL-2R chains.

Structural/biological feature	IL2R$_\alpha$	IL2R$_\beta$
Dissociation constant 10^{-9} mol l^{-1}		+
p55	+	
Intermediate affinity		+
Long cytoplasmic tail		+
Soluble form	+	
Probable signal transduction		+
Highest nonpeptide content	+	
Slow IL-2 association rate		+
Binds IL-2	+	+

6.7

Property	IL-1	IL-2
1) Macrophage-T interaction induces?	+	+ (1)
2) Bind receptors on T cells	+	+
3) Two species of the cytokine	+	
4) Hydrophobic leader sequence		+
5) Single receptor chain	+	
6) Clonal expansion of T cells		+ (2)
7) Endocrine activity	+	
8) Receptors on a majority of cells	+	

Comments: 1) This is equivalent to an APC-T interaction which first results in IL-1 production by the APC which then stimulates IL-2 production by the T cell. 2) Once activated IL-2 alone can support the clonal expansion of T cells.

6.8

1) True.

2) False. IL-2 is mainly produced by CD4$^+$ T cells although IFN-γ is produced by all T cells. Only IFN-γ is known to enhance MHC II expression on APCs.

3) False. Although IFN-γ is involved in activation of inflammatory cells and T and B cells it does not promote IgG and IgE class switching. It induces class switching to IgG2a.

4) False. IL-1β is not secreted as a 17kD product but requires extracellular processing (see Section 6.5.1).

5) True. IFN-γ produced by one cell acts on neighbouring cells.

6.9   This question should have helped you distinguish between these three cytokines.

Property/effect	IL-4	IL-5	IL-6
1) Induces class switching to IgG1 in mice	+		
2) Inbduces acute phase proteins			+
3) Promotes production of IgG1 and IgE	+		+ (1)
4) 3 intrachain S-S bonds	+		
5) B cell differentiation factor		? (mice)	+
6) Mast cell growth factor		+	
7) Macrophage activating factor		+	
8) Augments IL-2R in B cells		+	
9) Natural inhibitor	+ (2)		

Comments: 1. Notice, the question says promote not switch. IL-4 promotes production of both by inducing class switching. IL-6 as a differentiation factor induces plasma cell differentiation and therefore production of all Ig classes. 2. A soluble form of IL-4 may act as a natural inhibitor.

6.10   This was intended to give you some practice at interpreting data based on the activities of antibodies. If you have made all the correct conclusions you should congratulate yourself. If not, do not be too concerned as they were fairly difficult.

1) Incorrect conclusion. The anti-sheep IL-2 antiserum is not specific for the receptor but for the cytokine. It blocks the activity of IL-2S as it binds to the active site of the cytokine thereby inhibiting binding of the IL-2S to the receptor chains.

2) Correct conclusion. In the biological assay, you will notice that addition of the anti-IL-2H does not affect the growth of the cells. Since the antibodies do precipitate the affinity linked complexes, they are specific for the IL-2H but do not bind the active site of IL-2H and therefore do not affect the biological activity of the cytokine.

3) Incorrect conclusion. Although they do express IL-2 activity, the evidence does not support structural homology. You can conclude this from the observation that polyclonal antibodies directed against either cytokine do not cross react since no immunoprecipitation results except when using the homologous antibodies. This conclusion is also supported by the observation that anti-Tac does not immunoprecipitate complexes from IL-2S affinity linked sheep cells suggesting that the binding sites of the two cytokines are structurally different.

4) Correct conclusion. Anti-Tac, as you know, is a monoclonal antibody prepared against human IL-2α chains. The data shows that anti-Tac also binds a similar IL-2R chain on sheep cells.

    However, the affinity linking experiments indicates that IL-2S once bound to the IL-2R blocks binding by anti-Tac. This contrasts to the results using IL-2H which does not block subsequent binding by anti-Tac. This suggests that the binding of IL-2S differs from that of IL-2H, possibly binding to different sites. An alternative explanation is that the two cytokines bind the same receptor site but that IL-2S sterically hinders the binding of anti-Tac.

5) Correct conclusion. The presence of 5 bands on the autoradiogram following immunoprecipitation with anti-IL-2S compared with 3 receptor bands using anti-IL-2H precipitation of affinity linked IL-2H suggests (but does not prove) the presence of other proteins or cytokines in the crude supernatant which induced antibodies as well as the IL-2S. It is also possible, of course, that more sheep IL-2R components have been detected using IL-2S affinity linking since some components may not bind IL-2H. The correct approach would be to obtain a recombinant sheep IL-2 and use this in these experiments.

6.11   1) Activation of T cells and progression from **G0** to **G1** of the cell cycle involves crosslinking of **TCR - CD3** with antigenic **peptides** bound to **MHC II**, interactions between **cell adhesion molecules** of the T cell and APC, production of **IL-2** by the activated T cell and IL-1 by the APC.

2) **IL-6**, secreted by APC, promotes the expression of **high affinity** IL-2R on the T cell. **IL-2**, stimulated by the APC derived cytokine **IL-1**, promotes the progression of the T cell through the cell cycle when the T cell expresses receptors for **transferrin**.

3) IL-2 production is dependent *in vivo* on the continued presentation of **antigen** by **APC**. In the laboratory, T cell **proliferation** is antigen **independent** and IL-2 **dependent**.

6.12   You will already know the cell adhesion/activation molecules from the last chapter but this question helps to emphasise their role in cytokine action.

1) LFA-1 matches with ICAM-1 g). This is an important interaction during B cell activation.

2) IFN-γ matches with IgG2a f) since it promotes switching to this isotype.

3) IL-6 matches with B cell differentiation e) as it promotes development of plasma cells and Ig production.

4) IgE/IgG1 matches with IL-4b b) since the latter promotes class switching to these isotypes.

5) CD2 matches with LFA-3 a). Again highly avid interactions between these ligands promotes B cell activation.

6) TCR-CD3 matches with MHC II-peptide c). That was easy, wasn't it?

7) Inhibits Ig synthesis matches with TGF-β d).

6.13   This question should have helped you to distinguish between TH1 and TH2 and should have not caused any major problems.

Feature	TH1	TH2
1) IFN-γ production	+	
2) IL-4 & IL-5 production		+
3) DTH reactions	+	
4) Cytokines enhance MHC II expression	+	+
5) Induce mast cell mediators		+
6) IL-2	+	
7) IL-3	+	+
8) Promote IgE production		+
9) Activated by MHC-peptide	+	+
10) TNF-β production	+	

Responses to Chapter 7 SAQs

7.1   Set 1. The odd man out in this set is MHC II-foreign peptide (B) since this involves mature immunocompetent T cells which have passed through the thymus. The other 4 items refer to the development of T cells. Hence, the pre-T cells may have arisen in the foetal liver (C), then move to the thymus where they undergo DNA rearrangement (A) and express CD1 (E), only found on immature T cells. The immature T cells are then exposed to MHC on thymic epithelium (D) when they are positively selected.

Set 2. This in a way is the converse of set 1. The common element in four of the items is the immune response. Hence lymph nodes (D) are secondary lymphoid organs (E) where responses occur resulting in antibodies (A) and these responses involve T and B cells of the recirculatory pool (C). The thymus (B) is the odd man out as it is involved with development of T cells not with immune responses.

7.2   1) Incorrect. The network of fibres are reticulin not collagen.

2) Incorrect. Some of the follicles may be primary follicles which do not contain germinal centres. The latter are the sites for B cell-FDC interactions.

3) Correct.

4) Incorrect. Prominent cells in the medullary cords are plasma cells whilst in sinuses, the major cell type are the macrophages which capture antigen in lymph, especially if antibodies are present.

5) Incorrect. Lymph nodes are secondary lymphoid organs. These are defined as sites where T and B cells respond to antigen. Primary lymphoid organs are those involved in T and B cell development.

7.3   1) Incorrect. It has been estimated that about 10% of the lymphocytes found in the node are derived from afferent lymph; the remainder enter directly from the blood via the post-capillary venules.

2) Correct. Most of the antigen is captured and destroyed by cells such as phagocytic macrophages and takes no part in the stimulation of T and B cells.

3) Correct. If you look back to Chapters 5 and 6 you will realise that since T cells can be activated in both these sites, then cytokine production also occurs.

4) Correct. We have already said that antibodies pass out of the stimulated node via the efferent lymph. We think you will see that, since cytokines are produced in the node, any not used will find its way into efferent lymph. This has been shown experimentally.

5) Correct. Since recruitment involves greatly increased numbers of lymphocytes entering the node direct from the blood and only a small percentage of these will be retained by the node due to their antigen specificity, it should be obvious that the remainder of the cells will pass out of the node in efferent lymph on their way back to the blood.

7.4   The table should be completed as follows:

Activity	Lymph nodes	Spleen
1) Antibody production	Medulla	Red pulp
2) T cell priming	Paracortex	Periarteriolar sheath
3) Location of IDCs	Paracortex	Periarteriolar sheath
4) B cell-FDC interaction	Follicles	Follicles

Note: If you gave Cortex as the answer for 4) in lymph nodes then you are also correct. However, follicles are more accurate.

7.5   All of these items were concerned with haemopoiesis and you should have had no problem getting the correct matches.

1) B cells are matched with IL-7 C) as this cytokine appears to control early B cell development.

2) Granulocytes are matched with G-CSF A) as this cytokine promotes development of the granulocyte lineage.

3) T and B cells are matched with IL-3 B) as this cytokine promotes the early steps in development of the lymphoid lineage.

4) Macrophages are matched with M-CSF E) as this cytokine appears to promote development of monocytes from CFU-GM.

5) Red cells are matched with EPO D) or erythropoietin which is involved in the development of the burst forming units for erythrocytes from CFU-GEMM.

7.6   This was quite difficult but will help you to distinguish between these important cells.

Characteristic	Neutrophil	Basophil/ Mast cell	Eosinophil
1) Fc$_\gamma$RIII	+	+	+
2) Histamine production		+	
3) Binds C3b	+	+	+
4) High affinity Fc$_\varepsilon$R		+	
5) Vasodilation		+	
6) Low affinity Fc$_\varepsilon$R			+
7) Granules	+	+	+
8) Release of PAF		+	
9) Nonphagocytic		+	
10) Binds IgG opsonised antigens	+	+	+

Notes: since all 3 types of cells possess both Fc$_\gamma$RIII and CR1 we have to assume that they will bind IgG opsonised antigens 10) and C3b 3). Only basophils are known to be nonphagocytic although eosinophils are not very active either. Vasodilation, of course, results from histamine release by basophils/mast cells only.

7.7   This question was intended to demonstrate a typical experiment on selective migration of lymphocytes and at the same time improve your skills in interpreting results. If you performed well on this question, you can look forward with some confidence to becoming a good experimental immunologist!

1) Not a reasonable conclusion as antibodies A11 with specificity for HEBF$_{LN}$ exert no inhibitory effects on cells migrating to spleen (see Expt 2.E).

2) Not a reasonable conclusion since the results show that there may be more cells expressing HEBF$_{LN}$ rather than homing receptors (HEBF$_{PP}$) for Peyers patch. This can be concluded by comparing the greater number of cells failing to attach to cervical lymph nodes (Expt 1.A with A11 treatment) with that failing to attach to Peyers patch HEV (Expt 1.B with 1B2 treatment). Since A11 appears to inhibit a larger number of cells from attaching to HEV, we can conclude from the data there is a higher proportion of cells expressing HEBF$_{LN}$. For additional data you could also compare Expt 2.C using A11 and 2.D using 1B2.

3) This is a reasonable and correct conclusion since it can be assumed that anti-Ig antibodies would not block the interaction of the homing receptors with the HEV (Expt 1A and 1B using anti-Ig). This type of experiment, of course, represents a control showing that antibodies generally have no effect, only homing receptor specific antibodies have blocking activity.

4) This is a reasonable conclusion as Expt 2.E indicates that thoracic duct lymphocyte migration into spleen is not affected by either homing receptor antibodies.

5) This is a reasonable conclusion based on the observations that A11 blocks attachment of some cells to HEV of and prevents migration to cervical lymph nodes and 1B2 possesses similar characteristics regarding Peyers patch HEV.

7.8   Notice, in this case, you are told that only one statement is correct.

1) Incorrect. Mac-1 is not selective for monocytes, it also selects for neutrophils.

2) Correct. Since they are both members of the integrin family, they are structurally related. LFA-1, as you know from earlier material, possesses an α chain (CD11a) and a β chain (CD18). VLA-4/LPAM-1 also consists of an α chain (CD49d) and a β chain (CD29). The α chains in this family possess 25-65% homology and the β chains 37-45% homology.

3) Incorrect. As well as using interactions of cell adhesion molecules on the haemopoietic cells and the endothelium, you will remember that cytokines both activate the endothelium to upregulate molecules such as ICAM-1, ELAM-1 and VCAM-1 and that chemotactic cytokines such as IL-8 are involved in attracting the cells to attach to the endothelium and diapedese to the tissues.

4) Incorrect. It has not been established as yet that M-Ad is the target of VLA-4 or CD44. Hence the question mark in Figure 7.8.

5) Incorrect. Although its quite true to say that these three structures are related as members of the selectin family, it is only ELAM-1 which selectively binds to neutrophils. CD62 also binds to monocytes.

7.9   This was relatively easy compared to SAQ 7.8 and we are sure you had no difficulty although you will notice we are including material from earlier chapters.

Feature	IgG	IgM
1) Activates complement	+	+
2) Somatic hypermutation	+	
3) First detectable antibodies		+
4) Primary/secondary response differs	+	
5) Protective antibodies	+	
6) High affinity	+	
7) B cell receptor	+	+

7.10   1) Correct. Systemic anaphylaxis can be passively transferred by serum. This has been demonstrated by injecting unprimed guinea pigs with serum from a sensitised animal.

2) Incorrect. Degranulation does not result in platelet activation as this is mediated by PAF which is freshly synthesised and not stored in the granules.

3) Correct. Histamine release induces vasodilation which is seen as reddening of the skin (erythema); the loss of fluid from the blood vessels due to vasodilation results in accumulation of fluid in the tissues (oedema).

4) Correct. As you can see from Figure 7.12, all of these reagents stimulate mast cells.

5) Incorrect. Systemic anaphylaxis results from vasodilation and bronchoconstrictive effects of histamine.

7.11   This question should have helped you sort out the similarities and differences in the 3 types of immediate hypersensitivities.

Mechanism/condition	Type I hypersensitivity	Type II hypersensitivity	Type III hypersensitivity
Complement activation		+	+
Soluble immune complexes			+
Mast cells or basophils	+		+ (1)
Wheal and flare reaction	+		
Allergens	+		
Non-complement fixing Abs	+		(2)
Histamine	+		+
Neutrophil mediated damage	+	+	+ (3)
Bronchoconstriction	+		
Involves IgG/IgM		+	+
Red cell lysis		+	
Post-infection syndrome		+	+ (4)
Leukopenia		+	
Drug-induced	+	+	+ (5)

Comments: (1) Although you would probably associate mast cells/basophils with Type I, it is apparent that immune complexes can also induce the activation of basophils in capillaries. (2) This obviously refers to IgE which does not fix complement. IgG and IgM, both complement fixing antibodies, are involved in Types II and III hypersensitivities. (3) We think you would have implicated neutrophils in both Types I and III mechanisms. However, cytotoxic mechanisms may also involve neutrophils since they possess Fc receptors for IgG and they could phagocytose IgG opsonised cells. (4) You will have noticed that Types II and III often arise as a result of a previous infection. (5) Many drugs can induce all three types of hypersensitivity. Some drugs such as penicillin also induce cell mediated immune reactions (Type IV).

Responses to Chapter 8 SAQs

8.1   1) Correct. Since both CD4$^+$ and CD8$^+$ T cells possess CD2 they all bind sheep red cells and will appear in the pellet following gradient separation whereas the B cells will remain at the interface between the medium and the Ficoll.

2) Incorrect. Since the polymorphs are found in the pellet, they must have a density higher than that of the separation medium.

3) Incorrect. The antibodies would have blocked the formation of the rosettes and therefore the T cells would not have been separated from the B cells. However, anti-CD2 antibodies are mitogenic for T cells and thus the T cells would have proliferated.

4) Correct. PHA binds to TCR epitopes common to all T cells and the CD3 complex is found in all T cells. Therefore, both CD4$^+$ and Cd8$^+$ T cells would be activated.

5) Incorrect. Superantigens attach directly to MHC II molecules and this combination activates subpopulations of T cells.

8.2   This question should have sorted out in your mind which T cell populations are being activated in the MLR this being dependent on the alloantigens expressed.

1) Syngeneic MHC I and MHC II matches with MLR-, CML- D) since this is a syngeneic interaction resulting in self recognition.

2) Allogeneic MHC I, syngeneic MHC II matches with MLR+, CML+ C) as we found that even in the absence of CD4$^+$ T cell help (syngeneic at MHC II) CTL proliferation proceeds at a low rate and some cytolytic activity results.

3) Allogeneic MHC I, allogeneic MHC II matches with MLR++, CML++ A). These represent the optimal conditions for both MLR and CML.

4) Syngeneic MHC I, allogeneic MHC II matches with MLR++, CML- B). Although this will be a full blown proliferative response by CD4$^+$ T cells, there is no response by CTL.

8.3   The only incorrect statement is 4). When a cell suspension is prepared of a primary tumour it consists mainly of tumour cells. There is, therefore, a plentiful supply of tumour antigen which together with added IL-2 induce proliferation of tumour specific T cells.

Let us just go through the other statements as you may have found some rather difficult.

1) This is correct as you will remember from earlier work that each T cell only recognises a single MHC I or II haplotype. Since each individual possesses up to 6 MHC I antigens on each cell and quite a few MHC II antigens on APCs, many different alloreactive CD4$^+$ and CD8$^+$ T cells could be activated.

2) This is correct probably due to the fact that there is a much higher proportion of T cells responsive to alloantigens than to any single antigenic peptide. Additionally, the allogeneic cells may express a huge number of MHC molecules which may activate the alloreactive T cells whereas an APC may express a foreign peptide complexed to only a minor proportion of the total MHC molecules on the cell.

3) This is true for similar reasons as 2). Presentation of peptides derived from processed self MHC I molecules may be at the same level as foreign peptide presentation ie much lower than allogeneic molecules.

5) This is correct. The procedure follows closely the protocols of both the MLR and CML tests.

8.4   Option 5 is correct. We think you might have found this difficult and you should not be concerned if you made the wrong choice. You can congratulate yourself if you got it right for the correct reasons. Since it was obviously concerned with the classical developed type hypersenstivity (DTH) mechanism we expected you to choose IL-2 as the odd man out since all the others are involved in this mechanism. This does not mean that IL-2 could not be produced at this site but that it is not necessary for the DTH response.

8.5   All the items are involved in DTH reactions; however, only 1), 2) and 5) are involved in innate immunity since 3) and 4) involve antigen specificity. The answer therefore is 1), 2) and 5).

8.6   1) Correct. Listeriolysin facilitates the movement of *Listeria* through the endosomal membrane into the cytosol where some of the bacterial proteins will be subjected to proteolysis, attached to MHC I in the endoplasmic reticulum and presented on the cell surface to CD8+ T cells which will lyse the infected cell.

2) Incorrect. Although the DTH reaction often leads to the destruction of the pathogen, this is not always the result.

3) Incorrect. The CTL is MHC restricted and hence will recognise either MHC I (CD8$^+$ CTL) or MHC II (CD4$^+$ CTL), but not both.

4) Correct. Since the infected macrophage is being stimulated by a primed CD4$^+$ T cell, it can also act as an APC and stimulate the T cell into proliferation.

5) Incorrect. The statement is true except for the word solely. T cells at the site can also produce TNF.

8.7   This was probably more difficult than it at first appears. Do not be too concerned if you did not get it completely correct. The answer is 2) and 5) only. Exogenous antigens 2) are required by the CD4$^+$ T cells which promote the differentiation of CD8$^+$ T cells by producing cytokines 5). There is no requirement for a classical APC 1) (ie a cell constitutively expressing MHC II) because in many cases cells infected with virus or tumour cells do not normally express MHC II. Before differentiation can take place, therefore, the cell has to be activated by IFN-γ to express MHC II in order to activate the CD4$^+$ helper T cell. Type 1 interferons are not required for CTL differentiation 3). Lysis of the target cell 4) is obviously a result of the activity of a fully differentiated CTL rather than a requirement.

8.8

1) Incorrect. Although fully differentiated CD8$^+$ T cells are thought to produce low amounts of IL-2, the pre-CTLs which have exitted from the thymus produce either very low amounts of or no IL-2.

2) Incorrect. Present ideas implicate CD4$^+$ cytotoxic T cells as well.

3) Correct. One of the earliest events in cell to cell contact is linking of LFA-1 to ICAM-1; this obviously precedes lysis.

4) Incorrect. There is no reason why CTLs cannot kill CD4$^+$ T cells as the latter express MHC I.

5) Incorrect. Apart from the structural similarity between C9 and perforin, there is little similarity in the actual mechanisms.

8.9

This should have helped you to learn the common and different features of killing by these 3 cell types.

Anti-tumour mechanisms	NK	CTL	Macrophages
1) Phagocytosis			+
2) Perforin mediated lysis	+	+	
3) Lymphotoxin		+	+
4) Tumour specific mechanisms		+	
5) Antibody mediated killing	+		+ (1)
6) Extracellular killing	+	+	+ (2)
7) Complement mediated killing			+ (3)

Comment: (1) Antibodies are involved in ADCC by NK cells and Fc mediated killing by macrophages. (2) NK and CTL kill by an extracellular mechanism; macrophages can also kill this way by secretion of cytotoxins. (3) Macrophages can also kill C3b opsonised tumour cells.

8.10

This question should help you to distinguish between the mechanisms involved in Types I and IV hypersensitivities.

	Type I	Type IV
1) Named reaction	Anaphylactic	DTH
2) Mediated by	IgE on mast cells	Sensitised T cells
3) Time of reaction	Short (30 min)	2 days
4) Skin test result	Wheal and flare	Erythema and induration
5) Major histology	Oedema	Cell infiltrate
6) Passive transfer	IgE antibody	T cells
7) Effector molecules	Histamine, leukotrienes etc	Cytokines

8.11

This question was quite difficult but should have helped you understand experimental approaches to the identification of autoreactive T cells. Of the 5 answers, 2), 3) and 4) would provide good supporting evidence for T cell involvement. Let us just discuss each one in turn.

1) This would not be indicative of T cell mediated disease without further evidence because this information does not demonstrate that the T cells are responsible for the tissue damage. Other cells also present may be implicated.

2) This indicates that there are T cells present in the lesion which express specificity for MBP strongly suggesting these T cells are involved.

3) This again is good evidence that some CD4$^+$ T cells are responsible for the induction of the autoimmune disease. The anti-CD4 antibodies have probably eliminated the autoreactive cells.

4) As in 2), this is very good evidence to implicate T cells expressing specificity for a peptide derived from the self protein MBP in induction of the autoimmune disease. It suggests that the autoreactive T cells are being presented with the self peptide associated with MHC within the area of the lesion.

5) This does not provide any evidence for the presence of autoreactive T cells. It simply shows that T cells at the lesion are functional T cells.

8.12 This question gave you the opportunity to interpret the data. We are sure you appreciated the opportunity because if you succeeded, you are on your way to becoming a good researcher. The statements which are supported from the data are 3), 4) and 5). Let us examine each one in detail.

1) This conclusion may well be true but since antibodies were also produced to the TCR peptides, it is not an accurate one.

2) The protective mechanism does not appear to involve elimination of the encephalitogenic T cells since the data shows that T cells could be taken from the lymph nodes of protected animals which could induce the disease in naive animals.

3) Since the T cells specific for the TCR peptide could confer protection on other animals, this together with the conclusion in 2) suggests that suppressor cells are involved with specificity for the TCR idiotype.

4) The data clearly show that the T cells which confer protection to EAE are MHC I restricted. Note that we will discuss the mechanisms of this in the main text.

5) You could assume that since the T cells confering protection are MHC I restricted they will be CD8$^+$ T cells.

8.13 1) Incorrect. Chronic rejection is thought to have similar mechanisms including prolonged stimulation of CD4$^+$ T cells and macrophages leading to fibrosis. However, there is no formation of granulomas.

2) Incorrect. Sensitisation of CD4$^+$ T cells probably occurs in lymphoid tissue such as lymph nodes by either passenger leukocytes (leucocytes) of donor origin expressing alloantigens or soluble donor alloantigens.

3) Incorrect. You may say that even when antibodies are responsible for rejection that B cells need CD4$^+$ helper T cells to produce the antibodies. This may be for some of the antibodies such as IgG in acute rejection but anti-blood group antibodies, which are often involved in hyperacute rejection are mainly IgM and these may be produced without T cell help.

4) Incorrect. Hyperacute rejection appears to be due to antibodies binding mainly to blood group antigens on endothelial cells resulting in complement activation. This would result in complement mediated damage to endothelial cells and platelet activation resulting in thrombosis and blocking of blood vessels.

5) Correct. As with all other immune processes, cytokines have a central role to play in (i) activation of T cells to alloantigens, (ii) migration of sensitised cells to the transplanted organ and (iii) tissue damage to the donor tissues resulting in rejection.

Appendix 1

Major abbreviations used in the text

ADDCC = antibody dependent cellular cytotoxicity

APC= antigen presenting cell

$\beta_2 m = \beta_2$ - microglobin

B7 = B cell activation marker

BZ20 = see CD45

CAK = cytokine activated killer cell

C- = complement proteins (c1q, Clr, Cls, C2, C3, C4, C5, C6, C7, C8, C9)

C_α = constant domain of IgA heavy chain (similarly C_μ =IgM, C_ε =IgE C_δ =IgD and C_γ = IgG)

CD =cluster of differentiation or cluster determinant

CD2 = also called T11, LFA-2, Tp 50 and Leu 5; surface marker on T cells and Natural killer cells

CD4 = T helper cell marker

CD5 = marker found on all T cells (plus B cells which produce polyreactive antibodies)

CD8 = CTL (Tc) cell marker

CD11a = see LFA - 1

CD16 = FC γ R111 receptor; binds IgG1 and 1gG3

CD18 = see CR3 and LFA-1

CD19 = B cell marker, associated with 1gM

CD20 = B cell marker, related to Ca^{2+} uptake

CD21 = receptor for complement component C3d.

CD23 = IgE receptor, low affinity (also called Fc_ε R11)

CD26 = T cell surface component, interacts with intracellular matrix

CD28 = also known as Tp 44, expressed on all CD4 cells and 50% CD8 T cells; binds B cell activation marker B7

CDw32 = $Fc_\gamma R11$

CD35 = complement receptor for C3b (see CR1)

CD40 = B cell surface marker, may interact with FDCs

CD44 = homing receptor on T cells, binds to ligand on epithelium of capillaries

CD45 = leucocyte common antigen

CD45 = found on T cells (called T200) and B cells (B220); leukocyte common antigen

CD45R = isoform of CD45

CD45RO = isoform of CD45

CD54 = see ICAM 1

CD56 = major cell marker of Natural Killer cells

CD57 = major cell marker of Natural Killer cells

CD62 = endothelial marker - binds neutrophils and mononcytes (=PADGEM; GMP-140)

CD71 = transferrin

CD74 = membrane form of the invariant chain (IC)

CFU = colony forming unit (can be specified eg CFU-B = basophil colony forming unit,. CFU-Eo = eosinophil, CFU-GEMM = granulocyte, erythrocytes, monocytes, megakaryotes = myeloid stem cell, CFU-GM = granulocyte-macrophage)

C_H = constant domain of heavy chain

CMI = cell mediated immunity

CML = cell mediated lympholysis

ConA = concanualin A lectin

CR1 = complement receptor = CD35, binds C3b and C4b

CR3 = (also CD116/CD18) - binds bacteria and iC3b

CR4 = complement receptor - binds iC3b

CSF = colony stimulating factor eg GM-CSF = granulocyte macrophage CSF

CT6 = cytotoxic T cell line, totally dependent on IL-2

CTL = (or T_c) - cytotoxic T cell

DP = loci in H-21 complex in Man

DQ = loci in H-21 complex in Man

DR = loci in H-21 complex in Man

DX = loci in H-21 complex in Man

DZ = loci in H-21 complex in Man

EAE = experimental allergenic encephalomyelitis

ECF-A = eosinophil chemotactic factor of anaphylaxis

ELAM-1 = endothelial leukocyte adhesion molecules - 1

EPO = enythroietin

ER = endoplasmic reticulum

Fab = fragment antigen binding: comprised of V_H, C_H1, V_L, C_L domains; is monovalent

$F(ab')_2$ = as Fab but consisting of 2 Fab's joined by the hinge region, is divalent

Fc = fragment crystallisable. The COOH end of the antibody consisting of $C_H$constant regions (except CH1)

F_cR = receptors for the Fc region (also FCR)

$FC_\gamma R$ = receptor for IgG

$FC_\gamma R111$ = see CD16; binds IgG1 and IgG3

$FC_\varepsilon R$ = receptor for FC_ε

$FC_\varepsilon R1$ = high affinity receptor for IgE

$FC_\varepsilon R11$ = see CD23; IgE receptor, low affinity

FDC = follicular dendritic cell

GALT = gut associated lymphoid tissue

G-CSF = granulocyte colony stimulating factor

GM-CSF = granulocyte-macrophage colony stimulating factor

GMP-140 = see CD62

HEBF = high endothelial binding factor

HEL = hens egg white lysozyme

HEV = high endothelial cells in venules

HLA = major histocompatability gene complex in Man, human leucocyte antigen

HLA-A (B)(C) = MHC Class I gene in Man

H2 = mouse MHC

HRC = horse red blood cells

HSP = heat shock proteins (genes)

I-A = loci in H-2 complex in mouse

IC = invariant chain (see also CD74)

I-E = loci in H-2 complex in mouse

iC3b = degradative product of C3b

ICAM = intercellular adhesion molecule

ICAM1 = intercellular adhesion molecule-1; receptor for LFA-1 (also called CD54)

IFN = interferon eg IFN - γ = interferon - γ

Ig = Immunoglobin (1g1y, 1gG1 etc)

I_i = invariant chain

IL = interleukin eg IL-2 = interleukin 2

LAK = lymphokine activated killer cell

LAM-1 = leukocyte adhesion molecule 1

Leu 5 = see CD2

LFA = leukocyte function antigen

LFA-1 = leukocyte function associated antigen-1 (composed of CD11a and CD18- aids movement of T cells)

LFA-2 = see CD2

LFA-3 = leukocyte function - associated antigen 3

L-selectin = homing receptor on T and B cells, bind to ligands on epithelium of capelines (also called MEL-14)

LTC = leukotriene C (see SRS-A)

LTD = leukotriene D (see SRS-A)

Ly = markers of murine cells

Mac-1 = CD116/CD18 - an integrin

M-Ad = mucosal addressin

MALT = mucosally associated lymphoid tissue

MBP = myelin basic protein

M-cells = specialised epithelial cells lacking microvilli

MCP-1 = monocyte chemotactic protein

M-CSF = macrophage colony stimulating factor

MECA-79 = antibody binding to PN-Ad

MECA-89 = antibody binding to M-Ad

MEL-14 = see L selectin

MHC = major histocompatability complex

MIF = migration inhibition factor

MLR = mixed lymphocyte reaction

MS = multiple sclerosis

NCF-A = neutrophil chemotactic factor of anaphylaxis

NK = Natural Killer cell (also called large granular lymphocyte)

PADGEM = see CD62

PAF = platelet activating factor

PAS = periarteriolar sheath

PCV = post capillary venule

PHA = phytohaemagglutinin

PMA = phorbol myristic acetate

PN-Ad = peripheral nodes - addressin

RAG = recombinant activation genes

Slp = sex limited protein

SRC = sheep red blood cells

SRS-A = slow reacting substance of anaplylaxis (composed of LTC and LTD)

T11 = see CD2

T200 = see CD45

Tac = antigen associated with T cell activation

T_c = (or T_c) - cytotoxic T cell

TCR-1 = T cell receptor usually produced by CD4⁻ and CD8⁻ T cells (ϕ and δ chains)

TCR-2 = T cell receptor usually produced by CD4⁺ and CD8⁺ T cells (α and β chains)

TCGF = T cell growth factor (=IL-2)

TDL = thoracic duct lymphocytes

T_c = (or T_c) - cytotoxic T cell

TGf = tumour growth factors (α,β etc)

TH = T helper cell (also T_H)

T_H = T helper cell (also T_H)

TH1 = subset of CD4⁺ helper T cells (produces IFN-γ)

TH2 = subset of CD4⁺ helper T cells (produces IL-4, IL-5, IL-6, IL-10)

TIL = tumour infiltrating lymphocytes

Tla = thymus leukaemia antigen

TNF = tumour necrosis factor (α,β,etc); = lymphotoxins (α,β)

Tp44 = see CD28

TP50 = see CD2

TRAP = T cell receptor - associated protein

T_s =suppressed T cell

VLA = very late antigen

VLA-3 =produced by T_H cells, bind collagen

VLA-4 = produced by T_H cells, binds fibronectin (also VCAM on APCs)

VLA-5 = produced by T_H cells hind fibronetin

VLA-6 = produced by T_H cells binds laminin

Appendix 2

Summary of the properties and activities of major cell types involved in the immune system

Cell type	Principal markers*	Major activities	Secretions
T cells	TCR1 or TCR2; CD3, CD2, CD5 CD7. T_H express CD4, T_C and T_S express CD8	T_H regulates T/B cells and mediates CMI to non-viral pathogens	Cytokines eg IL-2-6 IL-9, 10 (TH2 only), IFN-γ, TNF, CSF's, TGF-β
	LFA-1/3	T_C kill virally infected cells and tumour cells	IL-2+/-, IFN-γ, TNF, GM-CSF, IL-3 + perforins
		T_S suppress T/B activities	?
B cells	Ig M/D receptor associated with other chains. Other isotypes on memory cells. MHC 11+, CD19, CD20, CD22. Most have $Fc_\gamma R11$ (CDw32) and $Fc_\epsilon R11$(CD23) and complement receptors CR1/2. LFA-1/3	Antibody production	Antibodies' cytokines IL-1α, TNFα, TGF-β
Macrophages	$Fc_\gamma R111/R11$ and complement receptors CR1/3 MHC 11+/- $Fc_\gamma R1$ (CD64) on all but tissue macrophages, the latter have CD71 (transferrin receptor) LFA-1/3	Major phagocyte. Can act as an antigen presenting cell. Breaks down and repairs tissues	Cytokines eg IL-1, TGF-β, TNF, IFNα//β, IL-6, IL-8 prostaglandins, clotting factors, complement components Cl-5 collagenase, hyaluronidase, angiogenesis factor
Natural killer cells (large granular lymphocytes)	CD2,$Fc_\gamma R111$, IL-2Rβ CR3, LFA-1/3 CD56,CD57 major markers	Cytolytic for some tumour cells and virally infected cells. Also performs antibody dependent cellular cytotoxicity through Fc R111	Some cytokines eg IL-2 and IL-3

Table A2.1 * All cells express MHC 1. The listing is not complete and does not include markers expressed on activated cells. In some instances, the markers are not found on all cells of that type.
The majority of the information given in the table is found in the text.

Cell type	Principal markers*	Major activities	Secretions
Interdigitating dendritic cells (IDC)	MHC II+, FcR-, CR-	Antigen presenting cell - primes T_H	Some cytokines eg IL-1, IL-6
Follicular dendritic cell (FDC)	MHC II-, FcR+, CR+	Trap antigen in the form of immune complexes, process it to iccosomes and pass these to B cells	?
Neutrophils	CD10 principal marker $Fc_\gamma R11$ (CDw32), $Fc_\gamma R111$ (CD16) CR1 (CD35), CR3 (CD11b) C5a receptor	Major phagocyte Phagocytoses IgG and C3b opsonised particles	
Basophils/ mast cells	$Fc_\gamma R11$, FcgR111 CR1, 3 and 4. $FC_\varepsilon R1$ (high affinity IgE receptor	Mediates Type 1 hypersensitivity mechanisms Nonphagocytic	Histamine, SRS-A (+ other leukotrienes) thromboxanes, prostaglandins, platelet activating factor, chemotactic factors NCF-A and ECF-A.

Mast cells also secrete cytokines eg TNF, IL-1, IL-3, IL-4, IL-5, IL-6 GM-CSF |
| **Eosinophils** | $Fc_\gamma R11$, $Fc_\gamma R111$, CR1, 3 and 4 $Fc_\varepsilon R11$ (low affinity IgE receptor (CD23)) | Regulates basophil/ mast cell activities. Promotes killing of IgG and IgE opsonised helminths | Histaminase, aryl sulphatase, phospholipase, major basic protein |

Table A2.1 (Continued) * All cells express MHC 1. The listing is not complete and does not include markers expressed on activated cells. In some instances, the markers are not found on all cells of that type. The majority of the information given in the table is found in the text.

Index

A